DRAFTING: METRIC

CHARLES ROHLMEIER

Drafting Department
C. A. Prosser Vocational High School,
Chicago, Illinois

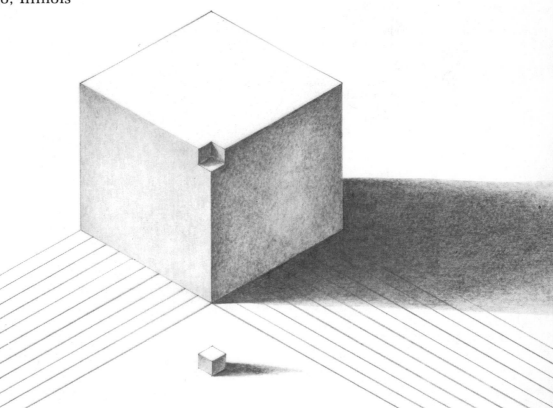

American Tech T.M.
American Technical Society   CHICAGO, ILL. 60637

DEDICATION

To Jeffrey and Bethany, my children, and to Sari, my wife.

PREFACE

The metric system is becoming the measuring system of the entire world. The nations of the world have begun an irreversible movement towards measurement standardization. Nowhere is this fact more evident than in industrial drafting rooms here in the United States. American industries that engage in foreign trade are already utilizing the metric system. Those who wish to sell goods in the future foreign market place must design and produce them with metric dimensions. Companies like Ford, General Motors, IBM, John Deere, and 3M are converting to metrics.

Auto mechanics need metric tools to repair foreign cars and many components of domestic cars. The pharmaceutical, photographic, and aerospace industries have been producing metric products for years. The language of science has been metric for some time.

The metric system is not hard to use or understand. In fact, it is much simpler than the conventional system. It has many advantages, but the most apparent are: (1) the use of a decimal system in which the use of complicated fractional units have been deleted, and (2) a standardized, single base unit of measure.

The metric system is nothing more than a simplified method of measuring things. The quantities do not change, only the names, and students of drafting need nothing more than the standard drafting instruments, a metric scale, and an open mind to prepare for a future in drafting.

DRAFTING: METRIC is a basic text composed of 22 simple units that break drafting down to its lowest common denominator. The units are arranged to provide continuous progress from the most basic manipulative skills through complex visualization concepts. The highly illustrated text is written in a clear, easily readable style that assures student success. It includes:

- Learning objectives
- New vocabulary terms
- Practice problems for immediate reinforcement
- Drawing problems related to industry
- A review of each unit
- Unit competency questions

DRAFTING: METRIC has been designed to help the beginning student of drafting meet the challenges of an all metric industrial world. All information is presented using only the metric system. Color is used to emphasize and highlight key information. An Appendix is included to serve as a review of basic math operations.

DRAFTING: METRIC is the answer to an effective, usable text for metric drafting.

The author wishes to express special thanks to R. Alfano and I. Kogan for inspiration; K. Gorski for numerous photographs; and the staff of the American Technical Society.

The Publisher

CONTENTS

THE METRIC SYSTEM & THE DRAFTER

This unit shows you how and why the drafter uses the metric system. You will discover:
- the beginning and growth of the metric system,
- the metric units of measure used on drawings.

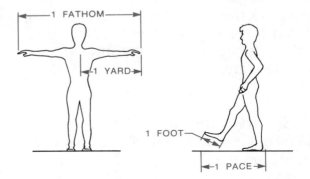

KEY WORDS

Metre: The basic unit of length in the metric system.

Millimetre: A unit of length equal to 1/1000 of a metre.

EARLY MEASUREMENT

In the early days of civilization, distances were often measured by comparing them to fingers, hands, feet, arms, and legs (Figure 1-1). This method was not accurate because not everyone was the same size (Figure 1-2).

As civilization advanced, methods of measuring advanced. It wasn't until the late 1700s that an accurate, easily understood system of measurement was developed.

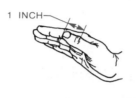

Figure 1-1
At one time fingers, hands, feet, arms, and legs were used for measurement.

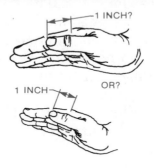

Figure 1-2
Measured lengths varied because people are not all the same size.

THE BEGINNING OF THE METRIC SYSTEM

In the late 1700s the French created a measuring system based on a unit of measure called the *metre*. The metre was determined to be the distance from the equator to the North Pole divided by 10 million (Figure 1-3). Based on this fixed distance, smaller and larger units of measure were determined (Figure 1-4).

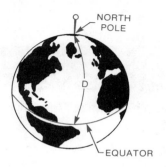

$$\frac{D}{10 \text{ MILLION}} = 1 \text{ METRE}$$

Figure 1-3
The metre was created by dividing the distance from the equator to the North Pole by ten million.

ACCEPTING THE METRIC SYSTEM

Use of the metric system for measurement spread throughout the world. One by one, countries abandoned their old measuring systems for the metric system. Today, only a few countries have yet to change to the metric system.

EVOLUTION OF THE MODERN METRIC SYSTEM

As technology increased, a need arose for extremely accurate measurements. The International Bureau of Weights and Measures developed le Systeme International d'Unites or the International System (S.I.). The International System set scientifically-based standards of measurement for the world to adopt and use. Today, the S.I. metric standards are the basis for modern drafting measurement (Figure 1-5).

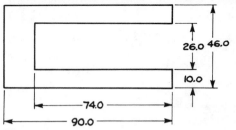

Figure 1-4
Larger and smaller distances are derived from the metre.

S. I. Metric Units

FOR	UNIT	SYMBOL
Length	Metre	m
Mass	Kilogram	kg
Time	Second	S
Electric Current	Ampere	A
Temperature	Kelvin*	K
Luminous Intensity	Candela	Cd
Amount of Substance	Mole	mol

*Kelvin used for scientific work, celsius used for non-scientific work.

Figure 1-5   S.I. metric units.

LENGTH MEASUREMENT

Linear measurement is most often used in the drafting room. Length is measured in *millimetres* on mechanical drawings (Figure 1-6). A millimetre is a unit of

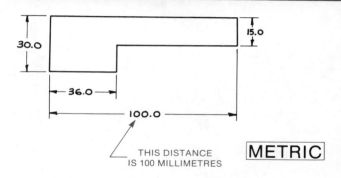

THIS DISTANCE IS 100 MILLIMETRES

METRIC

Figure 1-6   *A drawing with millimetres.*

length derived from the metre. There are one thousand millimetres in one metre (Figure 1-7).

USING THE MILLIMETRE

A dime is approximately a millimetre thick (Figure 1-8). The millimetre is a convenient size for making small or large measurements (Figures 1-9 and 1-10).

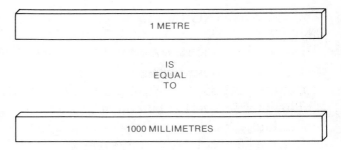

Figure 1-7
A metre is equal to 1000 millimetres.

Figure 1-8   *A dime is about 1.0 mm thick.*

| 1 METRE DIVIDED BY 10 = 1 DECIMETRE |
| 100 = 1 CENTIMETRE |
| 1000 = 1 MILLIMETRE |
| 1 METRE MULTIPLIED BY 10 = 1 DEKAMETRE |
| 100 = 1 HECTOMETRE |
| 1000 = 1 KILOMETRE |

Figure 1-9   *A drawing with small distances.*

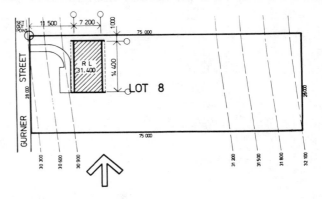

Figure 1-10
A drawing with large distances. (Standards Association of Australia)

- Early measurement was done by using fingers, hands, feet, arms, and legs.
- The metric system began in France in the late 1700s.
- Only a few countries have yet to change to the metric system.
- The S.I. metric standards are the basis for drafting room measurement.
- Measurements on mechanical drawings are done in millimetres.
- A millimetre is equal to .001 metre.

DRAFTING: YESTERDAY AND TODAY

This unit shows you how drafting has become the important job it is today. You will discover:
● drafting of the past compared to drafting today,
● the role of drafting in the world of work,
● the drafter's job and job future.

KEY WORDS
Drawing: A picture or diagram made of lines.
Drafting: A process for making accurate drawings of buildings and machines using special symbols and tools.
Drafter: A person trained in drafting techniques who makes drawings using drafting tools.

EARLY DRAWINGS
Since the early days of the human race, *drawing* has played an important role in telling thoughts and ideas. Prehistoric people communicated by sketching on the walls of caves (Figure 2-1).

As human knowledge developed, so did drawings. Ancient cultures drew plans to aid in building palaces and tombs (Figure 2-2). Some ancient drawings were made on a paper-like substance called *papyrus.*

Figure 2-1
Man has used sketching throughout history. (Field Museum of Natural History, Chicago, and sculptor, Frederick Blaschke)

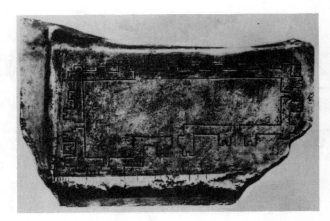

Figure 2-2
Ancient cultures used drawings to build palaces and tombs. (The Louvre, Paris, France)

As the world became more complex, drawings became more complicated as well. In the Enlightened Age, artists like Leonardo da Vinci recorded their thoughts, ideas, and designs in a drawing style similar to the style still used today (Figure 2-3).

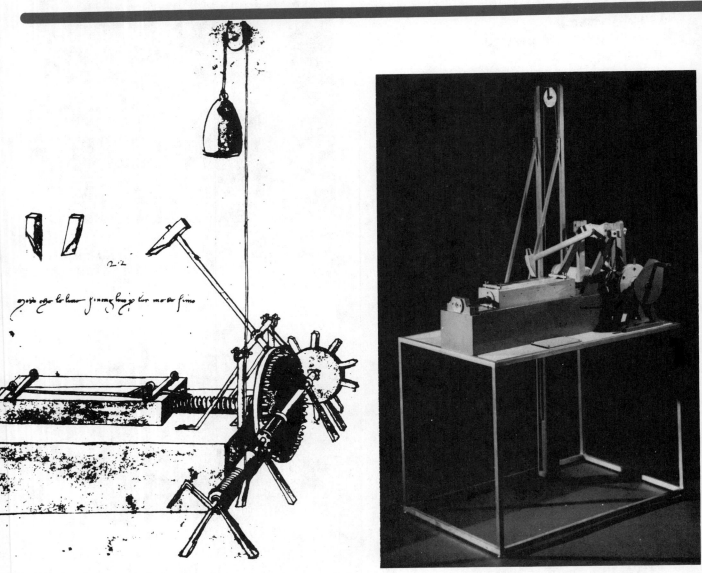

Figure 2-3
A sketch by Leonardo da Vinci of a device for file cutting. (Biblioteca Ambrosiana, Milan) Model made from this sketch is at right. (IBM)

As technical ability advanced, drawings became more and more complex. When North America was first being settled, complex drawings guided the construction of buildings still admired for their excellence today (Figure 2-4).

FROM EARLY DRAWING TO DRAFTING TODAY

In the Nineteenth Century, during the Industrial Revolution, manufacturing techniques using *drafting* and drawings were created (Figure 2-5). Drafting continued to grow with history: through the Steam Age, the Automobile Age, the Airplane Age, and into the Space Age.

Today, drafting is an important job. Thousands of people are employed world-wide to prepare drawings for almost everything that is made.

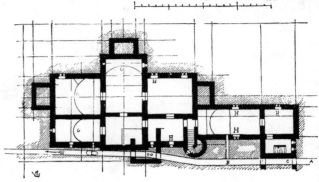

Figure 2-4
A complex drawing of an early Nineteenth Century building.

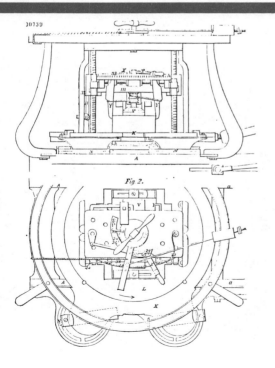

Figure 2-5
Veneering machine, 1854, side and top view, U.S. Patent No. 10739. Used for pressing and gluing thin wood strips on a wood base. (U.S. Patent Office)

THE NATURE OF AND NEED FOR DRAFTING

Every manufactured item we have or see today began as a drawing. In our highly technical world, the *drafter* and drafting play an important role in production. An automobile, for example, requires thousands upon thousands of drawings.

Each drawing (Figure 2-6) tells a complete story about the item it represents. This is known as the "Graphic Language" of drafting. With lines and symbols, the drafter draws a complete story of an item so that a worker, often many kilometres away, can know to the smallest detail exactly what the drafter has in mind.

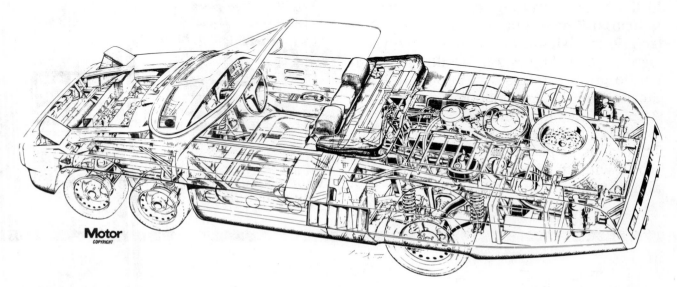

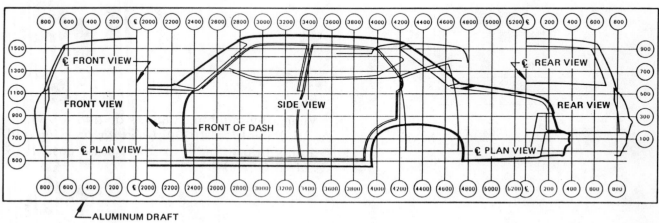

Figure 2-6
Top, *this automobile would require thousands of drawings.* (Motor Magazine, *England*)
Bottom, *this drawing is just one of many needed to produce an automobile.* (General Motors)

THE DRAFTING ROOM

Most drafting rooms have many people working to produce drawings (Figure 2-7). A typical drafting room has these people:

(Teledyne Post)

(E. I. duPont deNemours & Co.)

(Bruning Division, Addressograph Multigraph)

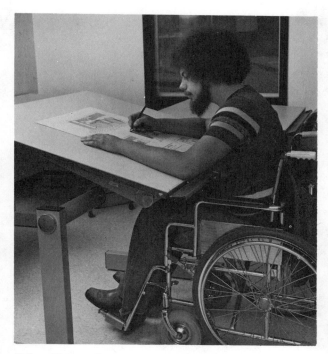

(The Huey Co.)

Figure 2-7
Many job opportunities exist in the field of drafting.

(Teledyne Post)

The *Chief Drafter*, in charge of all work done in a drafting room, must see that drawings are produced from the idea stage to a finished drawing.

Group Leaders or *Squad Leaders* are in charge of directing certain projects. Often many drafters are under the direction of one group or squad leader.

Drafters with much drawing experience often become *Designers*. They are responsible for creating drawings and calculations for new products.

Due to their many years of drawing experience, *Senior Drafters* draw the most advanced and difficult drawings in the drafting room.

Drafters are responsible for most of the drawings prepared. The drafter may someday become a senior drafter, designer, group leader, or chief drafter.

Junior Drafters are trainees who learn advanced skills while assisting the drafter and senior drafter.

Checkers see that the work done in the drafting room is free from mistakes.

The *Reproduction Technician* is responsible for making copies of drawings and technical material for all drafting room employees.

UNIT 2 REVIEW

- Prehistoric people used drawings to tell thoughts and ideas.
- Ancient civilizations used drawings to plan and build their cities.
- In the Enlightened Age, drawings became much more complicated.
- Drafting is a language of lines and symbols.
- Drafting was very important during the Industrial Revolution.
- Drafting was necessary in developing complicated items, such as automobiles, airplanes, and space vehicles.
- Drafting is an important, vital step in making almost everything today.
- Most drafting rooms in industry have drafters, junior and senior drafters, designers, checkers, group leaders, reproduction technicians, and a chief drafter.

DRAFTING EQUIPMENT

This unit shows you the drafting equipment you will need for a basic study of drafting. You will discover:
- the tools used for drafting,
- how to store and take care of drafting equipment.

KEY WORDS
Triangle: A three-sided figure.
Irregular: Not straight or even.
Scale: A measuring tool used by a drafter. A scale is similar to a ruler.

THE DRAFTER'S TOOLS
The correct tools are important for any job. A drafter, like other workers, uses special equipment.

BASIC EQUIPMENT
The following is a list of equipment needed for a basic study of drafting:
Drafting Board. A drafting board (Figure 3-1) supports the drafter's work. Boards are made in many sizes and styles. The size and style of board you buy is usually determined by the price you want to pay. A board 457 mm × 609 mm (18″ × 24″) is large enough for most student work.

Figure 3-1
Top, *wooden drafting boards. (Pickett Industries);* Bottom, *board with a metal edge. (The Gramercy Corp.)*

Drafting Tape. Drafting tape (Figure 3-2) is used to hold a drawing securely to a drafting board. When drafting tape is removed, it does not tear the sheet. A roll of tape 13 mm (1/2″) wide is normally used.

T-Square. A T-square is used mostly for drawing lines. T-squares come in many sizes and materials (Figure 3-3). When selecting a T-square, make sure the length of the blade suits the drafting board.

45° Triangle. The 45° triangle, (Figure 3-4) is a triangle used to draw lines. Plastic 45° triangles come in a variety of colors and sizes. A 203 mm (8″) 45° triangle is well suited for student work.

Figure 3-2
Drafting tape. (The Gramercy Corp.)

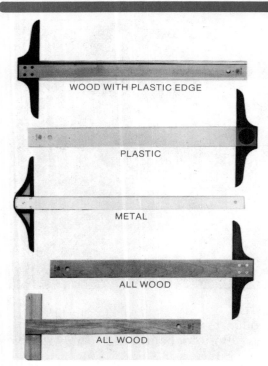

WOOD WITH PLASTIC EDGE

PLASTIC

METAL

ALL WOOD

ALL WOOD

Figure 3-3
Various types of T-squares. (Teledyne Post)

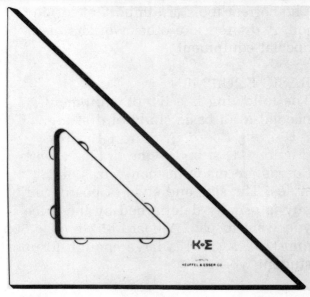

Figure 3-4
A 45° triangle. (Keuffel & Esser Co.)

30°-60° Triangle. There are various colors and sizes of 30°-60° triangles. A 254 mm (10″) 30°-60° triangle (Figure 3-5) is best for most drawing work.

Irregular Curve. Irregular curves (Figure 3-6) with many curved edges act as guides for drawing curved lines.

Lettering Guide. A plastic lettering guide (Figure 3-7) is used for making uniform letters and numbers.

Protractor. A protractor (Figure 3-8) is needed to measure angles.

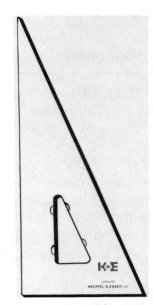

Figure 3-5
A 30°-60° triangle.
(Keuffel & Esser Co.)

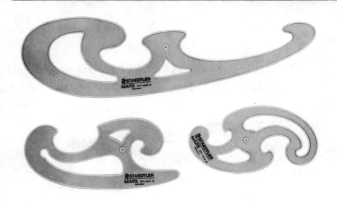

Figure 3-6
Irregular curves. (J. S. Staedtler Co.)

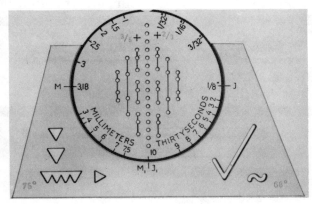

Figure 3-7
Guide for lettering. (Keuffel & Esser Co.)

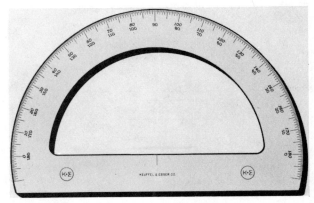

Figure 3-8
A plastic protractor. (Keuffel & Esser Co.)

Erasers. Many types of erasers are available (Figure 3-9). You should have at least one art gum type eraser for general erasing and one firm red or pink eraser for pencil erasing. Plastic erasers are used on film.

Erasing Shield. An erasing shield (Figure 3-10) helps erase small areas of a drawing. Shields are metal or plastic.

Bow Compass. A large bow compass (Figure 3-11) with a 150 mm opening is a good size for all circles you will need to make.

Dividers. A divider (Figure 3-12) with a 150 mm opening will handle the transferring and the dividing of distances.

Drafting Pencils. A good selection (several grades) of drafting pencils (Figure 3-13) should be purchased. Pencils are graded according to their hardness. H, 2H, and 3H grades are best for your use. Of these three, H is the softest. Still softer pencils range from HB through 7B; harder ones are numbered 4H through 9H.

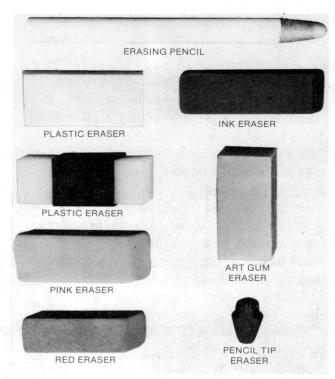

Figure 3-9
A variety of erasers. (Teledyne Post)

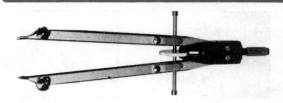

Figure 3-11
A bow compass. (Koh-I-Noor Rapidograph, Inc.)

Figure 3-12
Dividers. (J. S. Staedtler Co.)

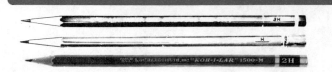

Figure 3-13
Drafting pencils of different hardnesses. (Top, Utley Co.; Bottom, Koh-I-Noor Rapidograph, Inc.)

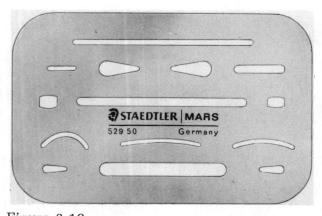

Figure 3-10
An erasing shield. (J. S. Staedtler Co.)

Lead Holder and Leads. If desired, a lead holder (Figure 3-14) can replace the drafting pencils. Lead holders require leads. Be sure to have enough H, 2H, and 3H leads.

Sandpaper Pad. A sandpaper pad (Figure 3-15) is used to point your pencils and compass lead.

Scale. A scale (Figure 3-16) is needed to make measurements. There are many styles of scales available. Any metric scale having the graduations 1:1, 1:2, 1:5, and 1:10 is needed. The 1:1 scale shows actual millimetres. One small division on the scale equals one millimetre.

Figure 3-14
A lead holder with drawing leads in various degrees of hardness. (Tru Point Products)

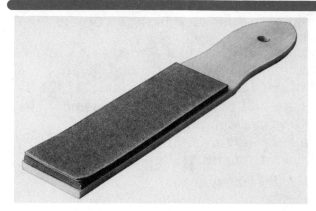

Drafter's Brush. A drafter's brush (Figure 3-17) is used to remove eraser particles and dirt from a drawing.

Drafting Powder. Drafting powder, which comes in a small can or dispenser (Figure 3-18), may be used to prepare and condition your drawing sheet.

Figure 3-15
A sandpaper pad. (J. S. Staedtler Co.)

Figure 3-16   *A drafting scale. (Utley Co.)*

Figure 3-17
A drafter's brush. (J. S. Staedtler Co.)

Figure 3-18
Drafting powder. (Keuffel & Esser Co.)

ADVANCED DRAFTING EQUIPMENT

Besides the basic equipment already mentioned, other optional equipment is also available. Optional equipment may replace some of the basic items. The following is a partial list of optional equipment for the student:

Drawing Table. A drawing table (Figure 3-19) may replace a drawing board.

Board and Parallel Straightedge. A drawing board with attached parallel straightedge (Figure 3-20) replaces the T-square and board as separate items.

Drafting Machine. The drafting machine (Figure 3-21) replaces the T-square, triangles (45° and 30°-60°), and the protractor.

Figure 3-19
A drawing table. (Teledyne Post)

Figure 3-20
Drawing board with parallel straightedge. (Keuffel & Esser Co.)

Figure 3-21
A drafting machine. (Keuffel & Esser Co.)

Lead Pointer. A lead pointer (Figure 3-22) replaces the sandpaper pad for pointing pencils. The pad is still needed for compass lead pointing.

CARE AND STORAGE OF EQUIPMENT

A set of drafting equipment represents a large investment. Caring for the equipment takes just an occasional few moments. Here are a few points to consider in caring for your equipment:

1) Wipe your equipment with a clean cloth before using.
2) Never use water to clean wooden items. They will warp if you do.
3) Avoid nicking or chipping the edges of your equipment.
4) Store your sandpaper pad in an envelope (Figure 3-23).
5) Store your equipment in an orderly manner. Never throw equipment.
6) Do not use your sandpaper pad over your drawing or your equipment.

Figure 3-22
A lead pointer. (Tru Point Products)

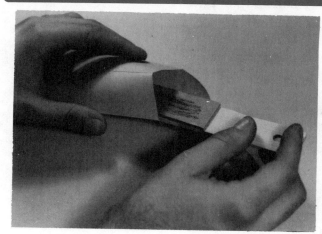

Figure 3-23
Keep your sandpaper pad in an envelope in order to keep your drawings and equipment clean.

- Basic drafting equipment includes:
  1 drafting board—457 mm × 609 mm
    (18″ × 24″)
  1 roll of drafting tape—13 mm (1/2″)
  1 T-square (to suit board)
  1 45° triangle—203 mm (8″)
  1 30°-60° triangle—245 mm (10″)
  1 irregular curve
  1 lettering guide
  1 protractor (large)
  2 erasers (1 art gum and 1 firm red)
  1 erasing shield
  1 bow compass (150 mm opening)
  1 dividers (150 mm opening)
  Several drafting pencils (H, 2H, 3H) or
    lead holders and leads
  1 sandpaper pad
  1 scale (metric)
  1 drafter's brush
  1 can of drafting powder
- Optional equipment may replace basic
    items.
- Keep equipment clean.
- Proper care of equipment is important.

DRAFTING MACHINES

This unit shows you how to use a drafting machine. You will discover:
- how to operate a drafting machine,
- how to care for a drafting machine.

KEY WORDS
Protractor Head: The main portion of a drafting machine used for controlling the machine.
Vernier: A scale used to locate an angle on a drafting machine.

DRAFTING MACHINES
There are many different types of drafting machines. All of them help the drafter in making drawings. The most common type of drafting machine is the track drafter (Figure 4-1).

THE PARTS OF A DRAFTING MACHINE
Figure 4-1 shows the main parts of a typical drafting machine. Each of these parts is important in operating a drafting machine.

Figure 4-1
A track drafter. (Mutoh Industry Ltd.)

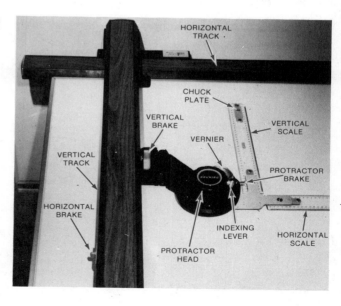

Figure 4-2
The important parts of a drafting machine.
(AM Bruning)

PREPARING THE DRAFTING MACHINE

Before a drafting machine can be used, several steps should be taken to prepare the machine. These steps are:

Step 1. The brake levers must be released. There are three brake levers: the horizontal brake, the vertical brake, and the protractor brake (Figure 4-3). When all three brake levers have been released, the drafting machine will move freely.

Step 2. The scales are carefully inserted into the head of the drafting machine. Place the scale flat on the board and firmly push the scale chuck plate into the chuck (Figure 4-4). Note: to remove the scale, use a scale wrench to push the scale out (Figure 4-5). Never pull a scale out. Besides damaging the scale or other equipment, you could injure yourself or someone else.

Step 3. Aligning the scale is a complicated procedure. Check with your teacher first.

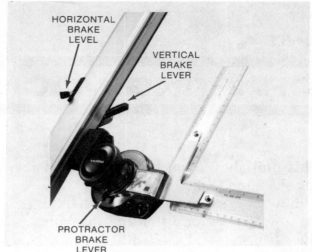

Figure 4-3   Brake levers. (Vemco)

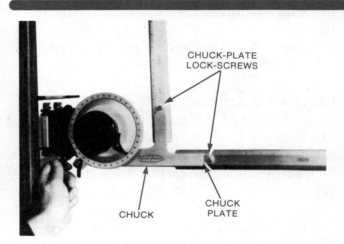

Figure 4-4   *Scales inserted into chuck. (Vemco)*

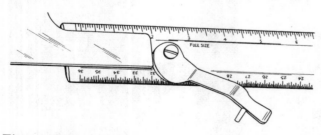

Figure 4-5
A scale wrench is used to remove a scale.
(Vemco)

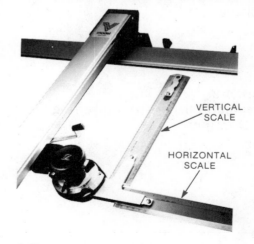

Figure 4-6
Horizontal and vertical scales. (Mutoh
Industry Ltd.)

OPERATING A DRAFTING MACHINE

The following are used when you operate a drafting machine:

A Horizontal Scale (Figure 4-6) is used for drawing and measuring horizontally.

A Vertical Scale (Figure 4-6) is used for drawing and measuring vertically.

A Protractor Head is used to set the angle for lines. Pressing the indexing lever (Figure 4-7) releases the head so that it can turn. By turning the protractor head, you can align the angle on the vernier (Figure 4-8). The protractor head stops and locks at each 15 degree mark. If you need an angle between the 15 degree marks, you can lock the protractor brake wing nut to hold the angle (Figure 4-7).

CLEANING THE DRAFTING MACHINE

When you clean a drafting machine, take care not to nick or scratch the scales. Wash them with soap and water. Wipe all other parts with a soft cloth.

Figure 4-7
Pressing the indexing lever. (AM Bruning)

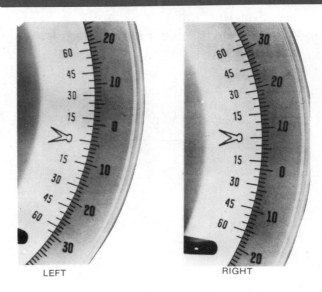

Figure 4-8
Left, 4½ degrees on a vernier; right, 8 degrees on a vernier. (Vemco)

- Drafting machines aid the drafter.
- The main parts of a drafting machine are: scales, protractor head, vernier, brake levers, indexing lever, and protractor brake wing nut.
- The brake levers should be released before aligning the scales.
- The scales should be aligned and then inserted or removed with care.
- The vertical and horizontal scales are used for drawing and measuring.
- Angles are set by pressing the indexing lever and turning the protractor head.
- The protractor head stops at each 15 degree mark. Other angles are held by locking the brake wing nut.
- A drafting machine should be cleaned carefully.

PENCIL PREPARATION

This unit shows you how to prepare your drafting pencil. You will discover:
- the different types of pencils,
- the correct pencils to use for drawing,
- how to prepare a pencil for drawing.

KEY WORDS
Pencil Grade: A system of rating the hardness of a pencil.
Lead: The writing part of a pencil.
Conical: Having the shape of a cone (Figure 5-1).
Graphite: A carbon formed with clay which is used for the lead portion of a pencil (Figure 5-2).

THE DRAFTING PENCIL
A drafting pencil is not an ordinary pencil. It has a smoother lead than an ordinary pencil. Because drafting requires fine, accurate lines, only a drafting pencil should be used.

Pencil Grades.   Pencils are graded according to their hardness. There are eighteen grades of pencils. A 9H pencil is the hardest and a 7B pencil is the softest. The grade for each pencil is stamped or printed at one end of the pencil (Figure 5-3).

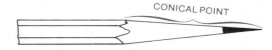

Figure 5-1
A cone. A pencil has a conical shape on tip.

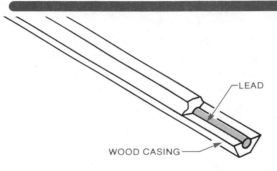

Figure 5-2   *Lead inside wood casing.*

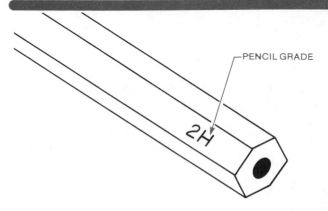

Figure 5-3
Common 2H pencil grade. Grade mark is stamped at one end only.

2B	SUPERIOR WRITING PENCIL
B	POPULAR WRITING PENCIL
HB	GENERAL PURPOSE PENCIL
F	
H	FOR MECHANICAL DRAWINGS
2H	
3H	FOR DISTINCT, SHARP OUTLINE DRAWING

Figure 5-4
Common pencil grades:
2B, B, HB, F, H, 2H, MEDIUM GRADE
and 3H. (J. S. Staedtler Co.)

A drafter chooses the grade of pencil suited to the drawing paper he or she will be using. Figure 5-4 shows the common pencil grades. For vellum and cross section paper the drafter uses pencils in the medium grade series. Hard pencils are used on layout paper. Soft pencils are not used on mechanical drawings because they smear easily.

Since the drawings in this text are designed for vellum, a 3H pencil should be used for light layout work; a 2H pencil for thin, dark lines; an H pencil for thick, dark lines; and an HB pencil for lettering work.

THE DRAFTER'S SHARPENER
In the drafting room, a pencil sharpener similar to a regular pencil sharpener is used (Figure 5-5). The drafter's sharpener removes only the wood casing of the pencil. Be sure not to remove the wood casing at the end stamped with the *pencil grade*. Note: do not sharpen both

Figure 5-5
Drafter's pencil sharpeners: Top, hand cranked. (Martin Instruments Co.); Bottom, electric. (Keuffel & Esser Co.)

ends of the pencil (Figure 5-6). This not only removes the grade marking, but it could be dangerous. All drafting pencils should have about 10.0 mm of wood casing removed before the *lead* is pointed (Figure 5-7). After the wood casing has been removed, the lead should be pointed.

THE SANDPAPER PAD

Use a sandpaper pad for *pointing* a pencil lead. The pencil is pointed by pulling it along the sandpaper pad (Figure 5-8). A twisting motion, while pulling, will give the lead an even *conical* point.

Always *point* your lead away from your drawing and equipment to keep loose graphite from falling on clean surfaces (Figure 5-9). Before using the pencil, wipe loose *graphite* from the point with a cloth or paper towel (Figure 5-10). Do not use ordinary drafting paper to wipe the point because it will leave loose graphite on the point.

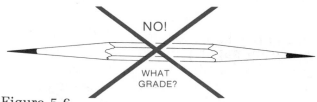

Figure 5-6
Do not sharpen both ends. The pencil grade is not known if both ends are sharpened.

Figure 5-7
A pencil with wood removed by sharpener.

Figure 5-8   *Point lead on a sandpaper pad.*

Figure 5-9
Point your pencil away from the drawing and equipment.

Figure 5-10
Wipe point with cloth or paper towel.

Sandpaper Pad Storage. Shake or tap your sandpaper pad over a wastepaper can to knock off the loose graphite (Figure 5-11). Store the sandpaper pad in an envelope to keep your other equipment clean.

LEAD HOLDERS, LEADS, AND POINTERS

If you use a lead holder and lead, the sandpaper pad can be used to point the lead. If you have a lead pointer, the sandpaper pad is not necessary. You can make a conical point by turning the lead and holder in the pointer (Figure 5-12).

Figure 5-11
Tap sandpaper pad on side of wastebasket to remove loose graphite.

Figure 5-12
The lead pointer replaces the sandpaper pad for pointing. (Keuffel & Esser Co.)

- Drafting pencils have smoother leads than ordinary pencils.
- Pencils are graded according to their hardness.
- Medium hard pencils are used for vellum and cross section paper.
- Hard pencils are used for layout paper.
- The wood casing of a pencil is removed with a drafter's sharpener.
- The pencil lead is pointed on a sandpaper pad.
- Leads in lead holders can be pointed by using a pointer.

UNIT 6

SKETCHING

This unit shows you how to sketch. You will discover:
- the basic steps of sketching,
- the uses of sketching to describe objects and shapes.

KEY WORDS

Sketch: A freehand drawing.
Pictorial Sketch: A sketch or drawing that shows the object in picture form.
Oblique: A form of pictorial sketching. Oblique sketching shows the front surface flat and the other surfaces on any convenient angle (Figure 6-1).
Isometric: A form of pictorial sketching. Isometric sketching is done on a 30 degree angle (Figure 6-2).

SKETCHING AND PLANNING

Before a drawing is started, you need to organize the drawing by making a sketch. A sketch is a quick method of organizing the parts of a drawing as well as the object itself being drawn (Figure 6-3).

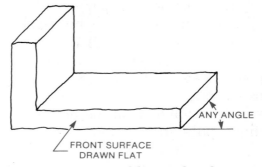

FRONT SURFACE
DRAWN FLAT

ANY ANGLE

Figure 6-1   *An oblique sketch.*

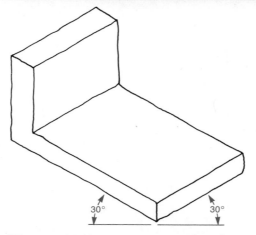

30°   30°

Figure 6-2   *An isometric sketch.*

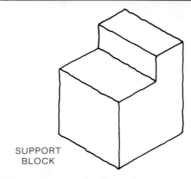

SUPPORT
BLOCK

Figure 6-3   *A sketch.*

Sketching helps determine where on the sheet the object should be drawn. Sketching also gives the drafter a clear view of how the complete drawing will look (Figure 6-4).

A sketch can be an object in a simple rough drawing (Figure 6-3). It can also be a complete drawing in rough form (Figure 6-4). Whatever form, the drafter should first know how to sketch.

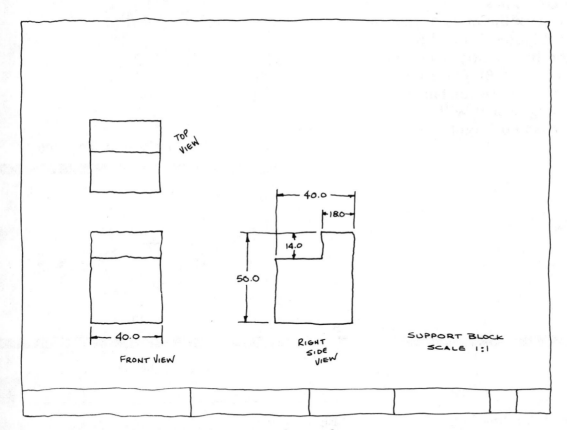

Figure 6-4   *A sketch of a drawing to be made.*

Tools Required for Sketching. Pencils of a suitable grade should be used for sketching. H and HB pencils are best for proper line weight. These pencils are soft enough to give the drafter free movement when pulling the pencil (Figure 6-5). If an eraser is needed, the drafter should use a soft eraser for neatness. Any paper is good for sketching.

Sketching Straight Lines. It is often helpful to first mark the ends of the line to be sketched (Figures 6-6 and 6-7). Then sketch the line lightly using broken short strokes (Figure 6-8). Once the line is sketched to the desired distance, darken it by going over it with more pressure on the pencil (Figure 6-9).

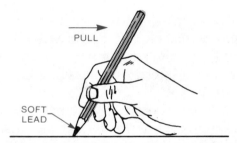

Figure 6-5 *Pull the pencil when sketching.*

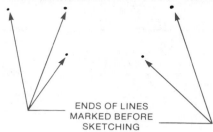

Figure 6-6
Mark the ends of the lines before sketching.

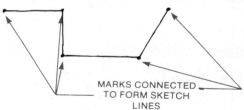

Figure 6-7
Sketch lines to connect the marks.

Figure 6-8 *Sketch the line lightly.*

Figure 6-9
Darken the line by using more pencil pressure.

Sketching Curves and Circles. To sketch a circle, first block out the area of the circle with a square (Figure 6-10). After determining the center point, (Figure 6-11), complete the circle by making sure the edge of the circle touches all four sides of the square (Figure 6-12).

Curves can be sketched easily. Mark the edges the curve will close off (Figure 6-13) and sketch the curve (Figure 6-14).

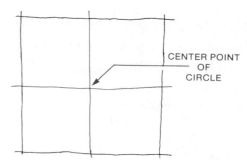

Figure 6-10
Block out the area of the circle.

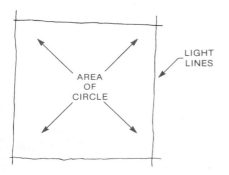

Figure 6-11 *Locate the center point.*

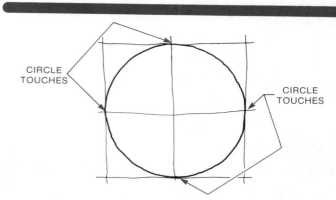

Figure 6-12 *Sketch the circle.*

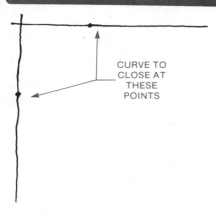

Figure 6-13
Mark the points the curve will close.

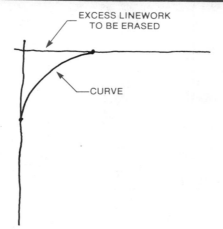

Figure 6-14 *The points closed by curve.*

Sketch the shapes shown. Be sure to
sketch each line lightly before darkening.

1)

3)

2)

4)

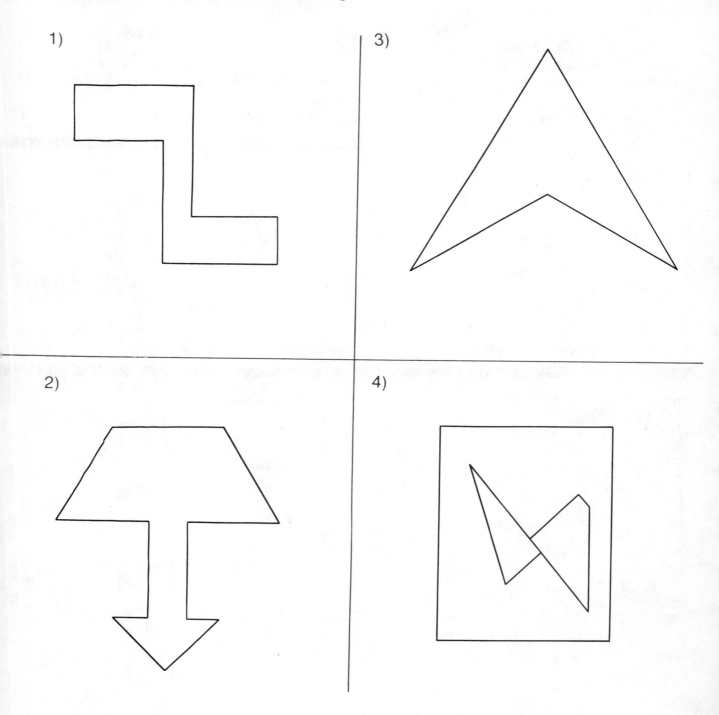

Sketch the shape shown. Be sure to block
out the area of the circle before sketching
it. The long and short dashed lines
(centerlines) indicate the center of the
circle and object.

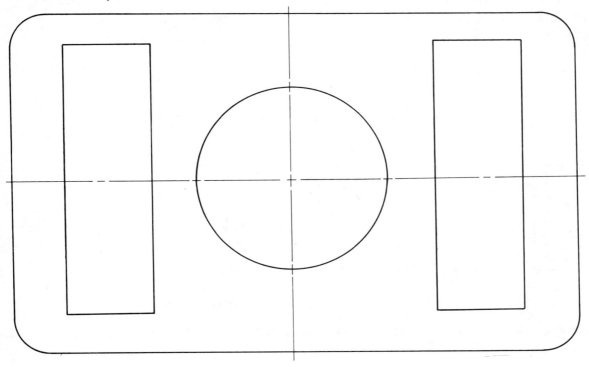

PICTORIAL SKETCHING

Two types of *pictorial sketches* are used for sketching: *oblique* and *isometric*.

OBLIQUE SKETCHING

Oblique sketches can take two forms: cavalier and cabinet. Figures 6-15 and 6-16 show a child's ordinary building

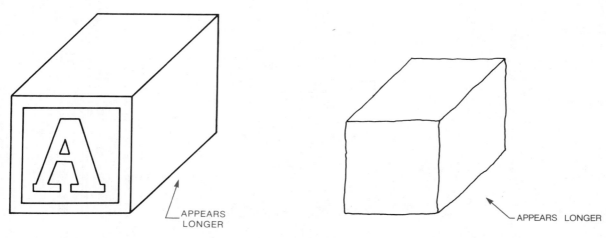

Figure 6-15   Left, *cavalier oblique drawing;* right, *cavalier oblique sketch.*

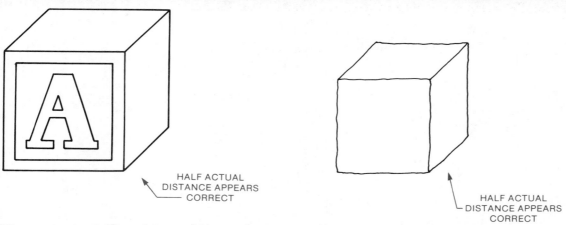

Figure 6-16   Left, *cabinet oblique drawing;* right, *cabinet oblique sketch.*

block. The cavalier oblique sketch (Figure 6-15) is rarely used because the object does *not appear* true in size. The cabinet oblique sketch (Figure 6-16) shortens the depth so that the sketch *appears more true* to form.

Cabinet Oblique Sketching. In the cabinet oblique method of sketching, the front surface is sketched flat (Figure 6-17). The front view appears as if it were directly in front of your face. Once the front surface is finished, the depth of the object is sketched with the edges leading back on any convenient angle (Figure 6-18). The depth is sketched half the actual distance (Figure 6-19). This makes the final sketch *look* true to size (Figure 6-20).

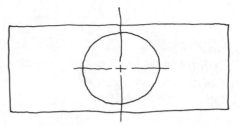

Figure 6-17   *Front surface sketched flat.*

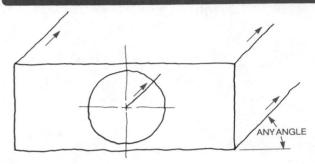

Figure 6-18
Edges lead back on approximate 45 degree angle.

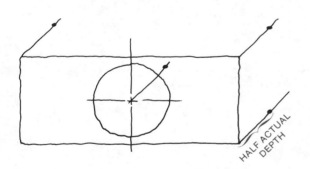

Figure 6-19   *Half the actual depth marked.*

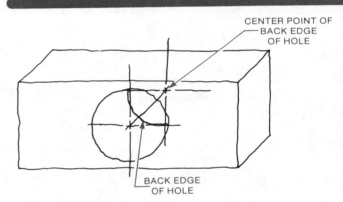

Figure 6-20   *Sketch closed off.*

Sketch the objects shown. Be sure to
sketch the depth to half of its actual
distance.

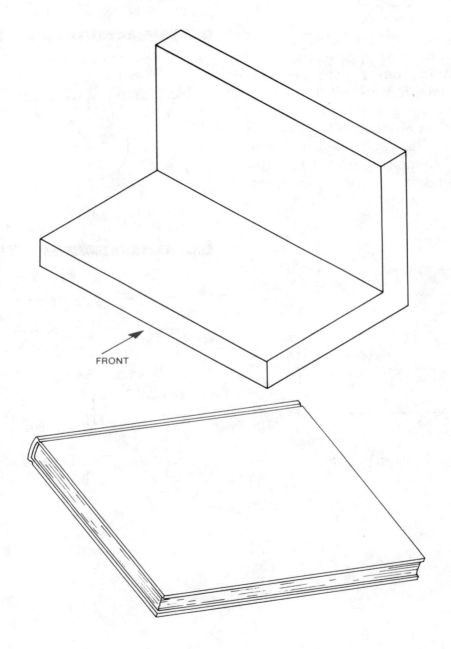

FRONT

ISOMETRIC SKETCHING

Unlike oblique sketching, an isometric sketch is done with all surfaces on angles (Figure 6-21). In isometric, the front surface is sketched on a 30 degree angle (Figure 6-22). After the front surface has been sketched, the edges are then sketched back on a 30 degree angle in the other direction (Figure 6-23). The actual depth of the object is then drawn and the figure closed off (Figure 6-24).

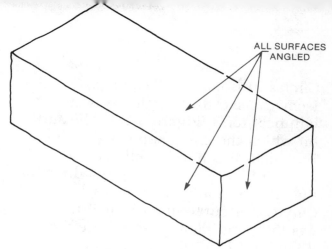

Figure 6-21   *All surfaces angled.*

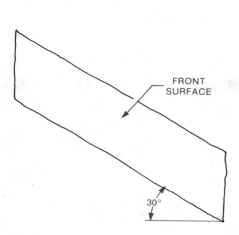

Figure 6-22
Front surface sketched on 30 degree angle.

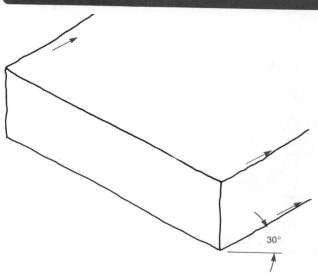

Figure 6-23
Edges sketched back on 30 degree angle.

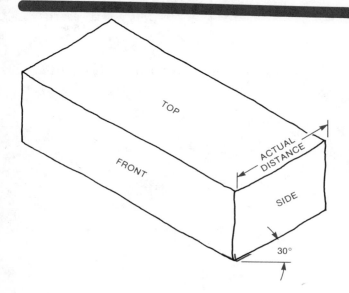

Figure 6-24
Actual depth marked. Object then closed off.

Circles in Isometric Sketching. On an isometric sketch, a circle takes an elliptical form (Figure 6-25). The surface on which the circle appears determines in which direction the ellipse slants. The cube in Figure 6-26 shows the direction the ellipse should slant for each surface.

Curves. A curve in an isometric sketch has the same slant as an ellipse does (Figure 6-27).

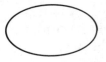

Figure 6-25 *An ellipse.*

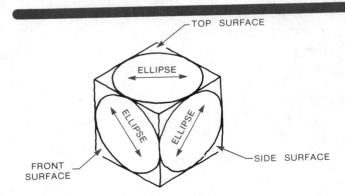

Figure 6-26
Isometric cube with elliptical circles.

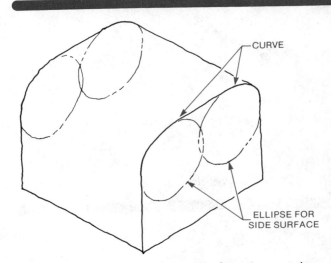

Figure 6-27 *Curves sketched in isometric.*

Make isometric sketches of the objects
shown. Review sketching ellipses (Figure
6-26) before starting.

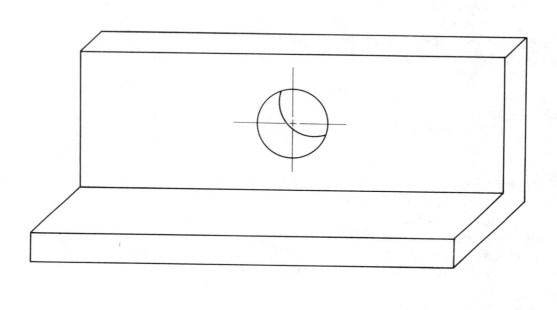

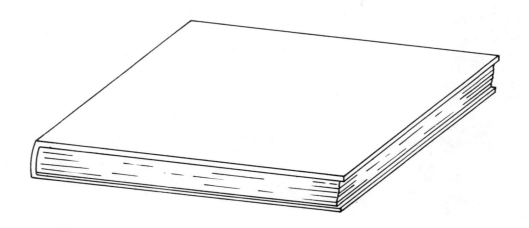

- Before drawing is started, a sketch should be made.
- Sketching helps the drawing layout.
- Sketching also plans the views of the object.
- A sketch can be a rough drawing of an object or a complete drawing in rough form.
- Soft pencils and soft erasers are used for sketching.
- It is helpful to sketch light lines and then darken them.
- Circles and curves should be "blocked in" before darkening.
- Two types of pictorial sketching are used: oblique and isometric.
- Oblique sketches are either cabinet oblique or cavalier oblique.
- The front surface of an oblique sketch is flat.
- The edges of oblique sketches go back on any convenient angle between 0 to 90 degrees.
- Isometric sketches are sketched on a 30 degree angle.

DRAFTING MEDIA

This unit shows you the drawing sheets used by drafters. You will discover:

- the types and sizes of drawing sheets,
- the sizes of borders, title blocks, and title strips,
- the uses of borders, title blocks, and title strips on your drawings.

KEY WORDS

Layout Paper: A heavyweight drawing paper used for accurate work.

Vellum: A tracing paper with a smooth surface used for drawings.

Print: A copy of an original drawing.

Film: A specially coated plastic drawing sheet.

Cross Section: Lines evenly spaced, running vertically and horizontally.

SHEET MATERIALS

Four types of drawing sheets are available: *layout paper, vellum, film,* and *cross section* paper. Each of these sheets has its own special quality.

Layout Paper. Layout paper is a heavyweight paper which is available in many grades, colors and surface textures. Layout paper is best suited for accurate drawings since it does not expand or shrink as much as other papers do.

Most new items drawn in a drafting room begin with a layout drawing. Layout drawings are extremely accurate drawings required in designing the exact size and shape of new items. Layout drawings are drawn on layout paper (Figure 7-1, *left*).

Vellum. Vellum is a heavy tracing paper. It has a fine surface texture suitable for pencil and ink line work (Figure 7-1, *center*). Vellum has a texture strong enough to withstand erasing without wearing away the sheet.

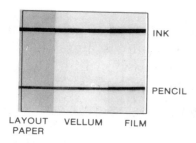

Figure 7-1
Types of drawing sheets: layout paper, vellum, and film. Note the difference in line qualities of pencil (bottom) and ink (top).

One of the most important features of vellum is that light shows through it. This enables you to trace other drawings through it without difficulty. Vellum is good for most *print* systems, which require that light be able to pass through (Figure 7-2).

Most industrial drawings are drawn on vellum. Student work is usually done on vellum since its cost is low compared to other drawing sheets.

Film. Film drawing surfaces are plastic sheets coated with a material that can be drawn on easily (Figure 7-1, *right*). This coating is called matting. Film, unlike paper, does not fold or tear easily (Figure 7-3).

Film drawings make better prints than vellum because film allows more light to pass through it. To draw on film, the drafter uses a pencil and eraser made

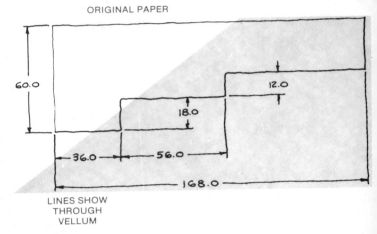

ORIGINAL PAPER

LINES SHOW
THROUGH
VELLUM

Figure 7-2
You can see through vellum. This helps in tracing.

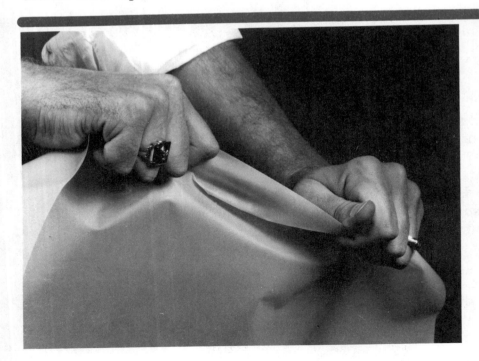

Figure 7-3
Film is tough.
(Keuffel & Esser Co.)

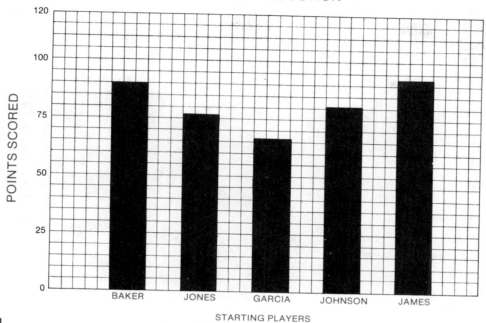

Figure 7-4
Cross section paper. The cross lines serve as guides. You can easily make graphs.

especially for film drawing. Because it is much more expensive than vellum, film is used mostly for professional work.
Cross Section Paper. Cross section paper is vellum with light cross section lines printed on it (Figure 7-4). The lines are equally spaced to help in laying out graphs and charts.

Cross section paper is often used for sketching before a formal drawing is made (Figure 7-5). Some cross section papers have lines that will not show on prints made.

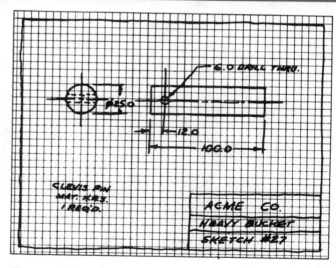

Figure 7-5
The lines in cross section paper help in quick sketching.

DRAWING SHEET SIZES

When selecting a sheet size, you must consider the size of the object to be drawn. Knowing the sheet sizes and how they compare to one another will help you avoid using a sheet too small or too large.

Drawing sheet sizes have been made uniform by drafting associations. These drawing sheet sizes are relative to one

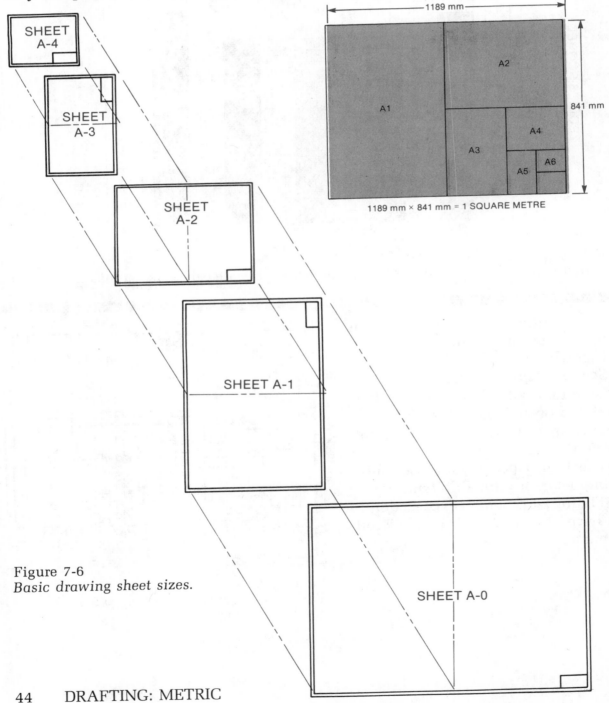

1189 mm × 841 mm = 1 SQUARE METRE

Figure 7-6
Basic drawing sheet sizes.

another (Figure 7-6). The chart below will
help you understand the sheet sizes and
their relationships:

Sheet Name	Size	Sheet Area Equals
A0	841 mm × 1189 mm	Sixteen A4 Sheets
A1	594 mm × 841 mm	Eight A4 Sheets
A2	420 mm × 594 mm	Four A4 Sheets
A3	297 mm × 420 mm	Two A4 Sheets
A4	210 mm × 297 mm	————

The Inch-Based Sheet Sizes. Before
adopting the metric system of
measurements, companies and schools
have used one of two drawing sheet
sizing systems. Some of these sheet sizes
are still used today.

The 8.50 × 11.00 inch (215.9 mm × 279.4 mm) Based System:

Letter Name	Sheet Dimensions
A	8.50 × 11.00 inches (215.9 mm × 279.4 mm)
B	11.00 × 17.00 inches (279.4 mm × 431.8 mm)
C	17.00 × 22.00 inches (431.8 mm × 558.8 mm)
D	22.00 × 34.00 inches (558.8 mm × 863.6 mm)
E	34.00 × 44.00 inches (863.6 mm × 1117.6 mm)

The 9.00 × 12.00 inch (228.6 mm × 304.8 mm) Based System:

Letter Name	Sheet Dimensions
A	9.00 × 12.00 inches (228.6 mm × 304.8 mm)
B	12.00 × 18.00 inches (304.8 mm × 457.2 mm)
C	18.00 × 24.00 inches (457.2 mm × 609.6 mm)
D	24.00 × 36.00 inches (609.6 mm × 914.4 mm)
E	36.00 × 48.00 inches (914.4 mm × 1219.2 mm)

Sheet sizes for most drawings in this text
are given with the drawing instructions.

BORDERS

Borders "frame" the drawing sheet. Border sizes vary. The borders shown in Figure 7-7 will be a guide for border sizes on drawings given in this text.

TITLE BLOCKS AND TITLE STRIPS

Title blocks and title strips are used on drawings for entering company or school names, dates, drawing numbers, and drafters' names (Figure 7-8). Title blocks and title strips vary from company to company and from school to school. The title blocks and title strips shown in Figures 7-9 and 7-10 will be used for drawings in this text.

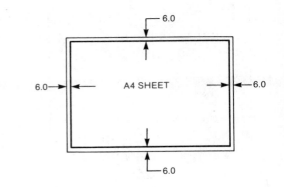

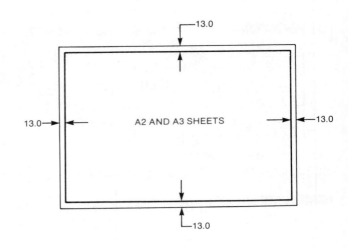

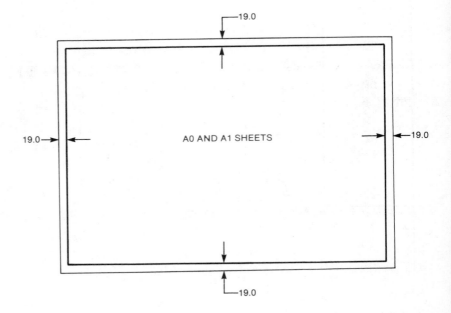

Figure 7-7 Sample borders.

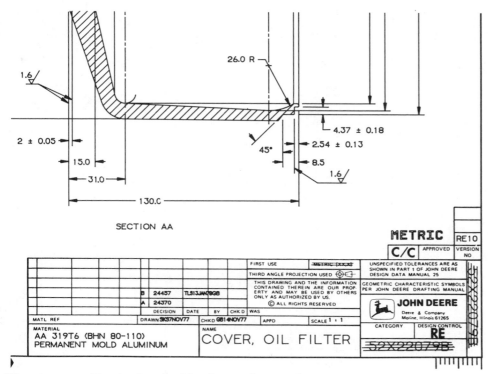

Figure 7-8 Typical title block used in industry. (Deere & Co.)

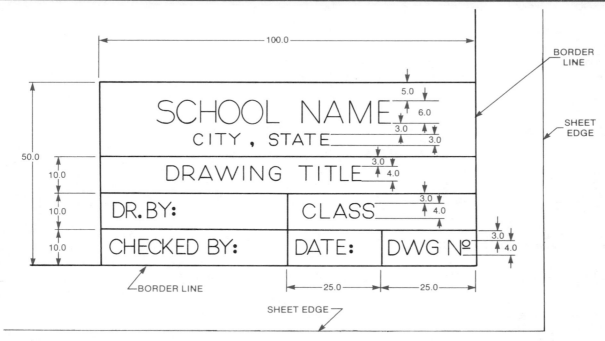

Figure 7-9 Sample title block.

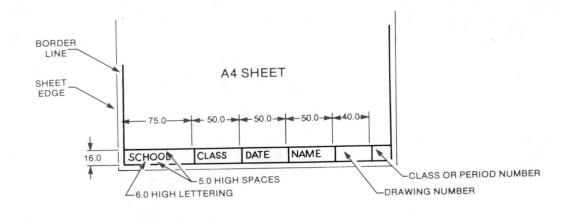

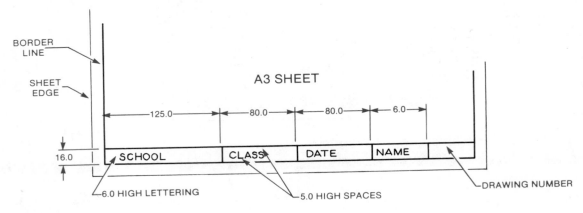

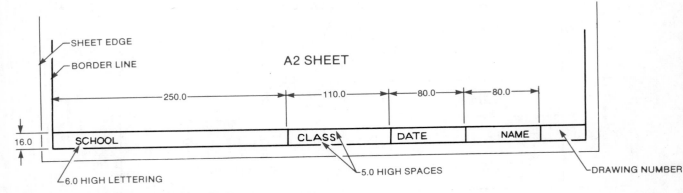

Figure 7-10 *Sample title strips.*

- Layout paper, vellum, film, and cross section paper are the main types of drawing sheets.
- Drawings of new items are first made on layout paper.
- Vellum has a good surface for drawing and erasing.
- Vellum allows light to pass through and allows quality prints to be made.
- Film is a plastic sheet with a matte surface.
- Film allows more light to pass through than vellum.
- Cross section paper is used for graphs, charts, and sketching.
- The standard sheet sizes for drawings are:

A0	841 mm ×	1189 mm
A1	594 mm ×	841 mm
A2	420 mm ×	594 mm
A3	297 mm ×	420 mm
A4	210 mm ×	297 mm

- Borders and margins "frame" a drawing sheet.
- Title blocks and title strips are used for company or school names, dates, drawing numbers, and drafters' names.

SHEET MOUNTING

This unit shows you how to mount drafting sheets on your board. You will discover:
- how to cover a drafting board,
- the steps in mounting a drawing sheet,
- how to condition a drawing sheet.

KEY WORDS
Linoleum: A hard, smooth covering with a canvas backing.
Vinyl: A plastic material that can easily be wiped clean.
Mounting: Securing a drawing in place.
Drawing Conditioning: Preparing a drawing sheet with drafting powder.

BOARD COVERINGS
Drafters use many different coverings to prevent holes and grooves from ruining the surface of the drafting board. Most drafting equipment manufacturers make *linoleum* and *vinyl* board coverings. Often, a sheet of paper with a hard surface is a suitable board covering.
Covering the Drafting Board.   Be sure to cover the entire surface of your drafting board without overlapping the edges (Figure 8-1). Use double-faced tape or an adhesive to apply a linoleum or vinyl covering (Figures 8-2 and 8-3). These board coverings can be cleaned by wiping with a damp cloth.

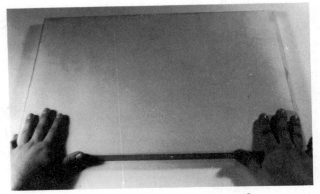

Figure 8-1   *Cover the entire board.*

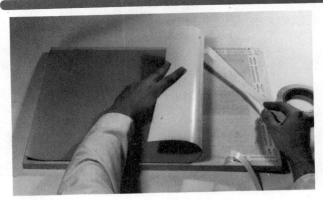

Figure 8-2   *Using double-faced tape.*

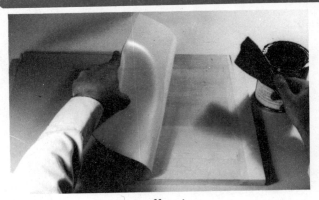

Figure 8-3   *Using adhesive.*

If you use a hard surfaced paper instead, tape down the corners of the sheet with drafting tape (Figure 8-4). Replace the sheet when it becomes dirty.

The board is now ready for sheet *mounting*.

MOUNTING A DRAWING SHEET

Place your drawing sheet to the left of the center of your board if you are right-handed and to the right if you are left-handed (Figure 8-5). Align the sheet exactly.

When borders *are not* printed on your sheet:

Step 1. Be certain the head of your T-square is firmly against the side of the board (Figure 8-6).

Figure 8-4   *Taping down corners.*

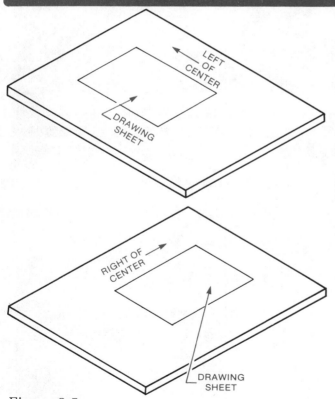

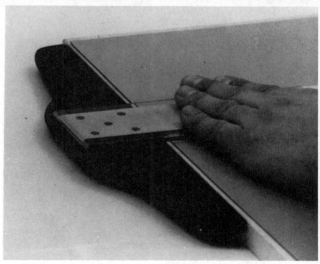

Figure 8-6   *T-square against board.*

Figure 8-5
Sheet in center of board.

Step 2. Align the top edge of your sheet with the top edge of the T-square (Figure 8-7).

Step 3. Smooth down the sheet while holding it in place (Figure 8-8).

Step 4. With two small pieces of drafting tape, tape down the two bottom corners of the sheet (Figure 8-9).

Step 5. Slide the T-square away, and tape the top two corners of the sheet (Figure 8-10).

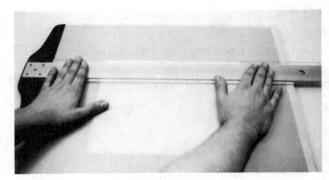

Figure 8-7
T-square and top edge of sheet aligned.

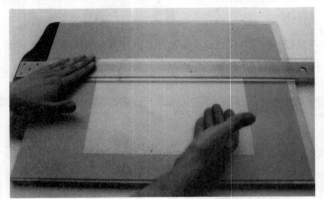

Figure 8-8   *Smooth down sheet.*

Figure 8-9   *Tape down bottom corners.*

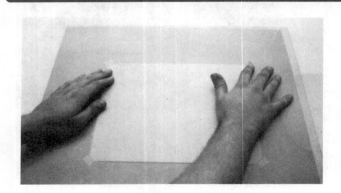

Figure 8-10   *Tape top corners.*

When borders *are* printed on your sheet:

Step 1. Hold your T-square firmly in place.

Step 2. Align the top border line of your sheet with the top edge of the T-square (Figure 8-11).

Steps 3-5. Same as when borders are not printed.

After the sheet has been mounted, it is ready to be conditioned.

CONDITIONING A DRAWING SHEET

On any drawing sheet (vellum, film, or layout paper) spread a powder (Figure 8-12) and rub it lightly on the sheet (Figure 8-13) before you begin drawing. The powder will keep your sheet free from dirt and grease. Many powders are available for treating a sheet before drawing.

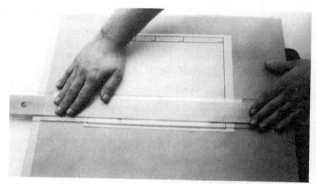

Figure 8-11
T-square and top border line aligned.

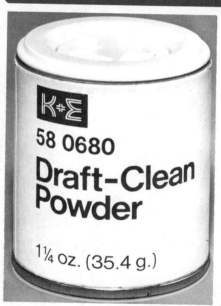

Figure 8-12
Drafting powder. (Keuffel & Esser Co.)

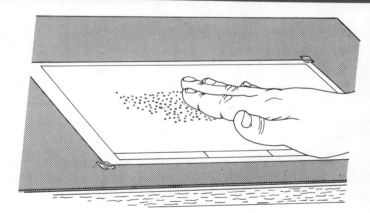

Figure 8-13   *Rubbing powder on.*

- Board coverings help protect boards from holes.
- Linoleum and vinyl board coverings are applied with adhesive or double-faced tape.
- Paper board coverings are taped to the board with drafting tape.
- Drawing sheets should be placed either to the right or left of the center of the board.
- Sheets without borders are aligned by placing the top edge of the sheet along the top edge of the T-square.
- Bordered sheets are aligned by positioning the top border line along the top edge of the T-square.
- The four corners of a drawing sheet should be taped down.
- Conditioning a sheet helps keep dirt and grease from your drawing.

HORIZONTAL AND VERTICAL LINES

This unit shows you how to draw straight lines. You will discover:

- how to use a T-square, a pencil, and triangles for drawing horizontal and vertical lines,
- how to draw accurate lines that are uniform in thickness and darkness.

KEY WORDS

Horizontal: Across the sheet; from side to side (Figure 9-1).
Parallel: Lines leading in the same direction and at the same distance apart at every point (Figure 9-1).
Tilt: To make something lean on an angle.
Rotate: To turn around.
Vertical: Straight up and down; at 90° to the base (Figure 9-1).
Slant: To lean; on an angle; a line between horizontal and vertical.

PREPARATION BEFORE DRAWING

Before any lines are drawn, the drawing sheet should be mounted and the surface conditioned as described in Unit 8. Then select a pencil of a suitable grade. Sharpen, point, and clean the pencil. (See Unit 5.) You are now ready to draw.

DRAWING LINES

Lines on any drawing should be straight, dark, and even. Whether the line is horizontal or vertical, there should be little difference in the appearance of the lines drawn (Figure 9-2).

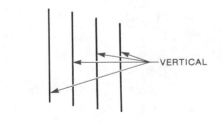
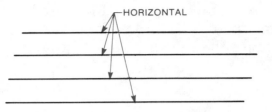

Figure 9-1
Parallel lines, vertical and horizontal.

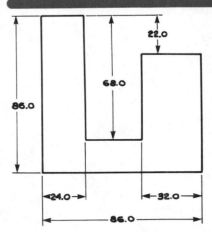

Figure 9-2 *Typical line work on a drawing.*

LINE WEIGHTS

Lines are drawn in three different thicknesses or weights (Figure 9-3).

Construction Lines are very light and thin; they can be erased easily. Construction lines are drawn to lay out each feature of a drawing.

Object Lines are dark and medium in thickness. They show the visible edges of the object being drawn. Object lines are about twice the thickness of construction lines.

Border Lines begin as construction lines. They are darkened, thicker lines used to frame the drawing. Border lines are three times the thickness of construction lines.

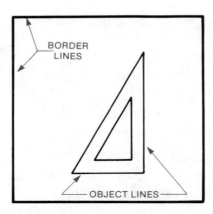

BORDER LINE

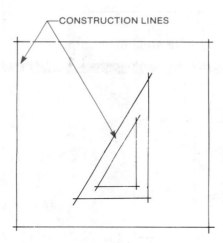

CONSTRUCTION LINES
ARE LIGHT AND THIN

OBJECT LINES ARE
DARK AND MEDIUM THICK

Figure 9-3 *Line thickness.*

DRAWING HORIZONTAL LINES

Horizontal lines are drawn by pulling a pencil along the upper edge of a T-square. The head of the T-square should always be placed on the left edge of the drawing board for right-handed drafters and on the right edge of the board for left-handed drafters (Figure 9-4). Hold the T-square firmly against the edge of the board with your free hand while you are drawing (Figure 9-5). This will keep horizontal lines parallel to each other and to the top and bottom of your drawing sheet. Use your free hand to move the T-square to the desired position.

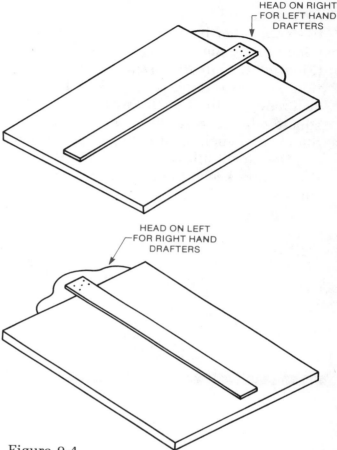

HEAD ON RIGHT
FOR LEFT HAND
DRAFTERS

HEAD ON LEFT
FOR RIGHT HAND
DRAFTERS

Figure 9-4
Top, *T-square position for left-handers;*
bottom, *T-square position for right-handers.*

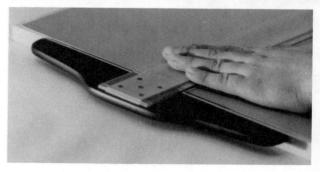

Figure 9-5   *Hold the T-square firm.*

While holding the T-square in place, use your other hand to pull the pointed pencil along the top edge of the T-square (Figure 9-6). *Tilt* the pencil in the direction you are pulling (Figure 9-7). For accuracy, tilt the pencil away from the T-square edge (Figure 9-8). While pulling, *rotate* the pencil with your fingers. Rotating the pencil allows the point to wear evenly and keeps the line uniform in thickness.

Figure 9-6
Pull the pencil along the top of the T-square.

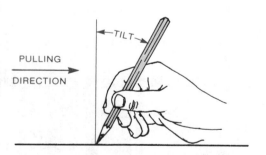

Figure 9-7   *Tilt the pencil while pulling.*

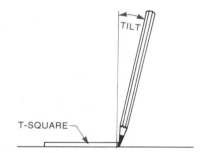

Figure 9-8
Tilt the pencil away from the T-square edge and rotate it.

PRACTICE SHEET 9-A PRACTICING HORIZONTAL OBJECT LINES

Step 1. Draw ten parallel construction lines horizontally (Figure 9-9). Be sure to hold the head of the T-square firmly against the side of your drawing board. Rotate the pencil. Space your lines so they will fill the entire sheet.

Step 2. Darken each of the ten construction lines (Figure 9-10) to the width of an object line (Figure 9-3). Point your pencil after each line if needed.

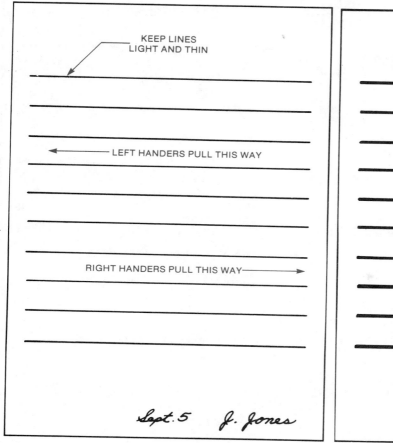

Figure 9-9
Step 1, *horizontal construction lines.*

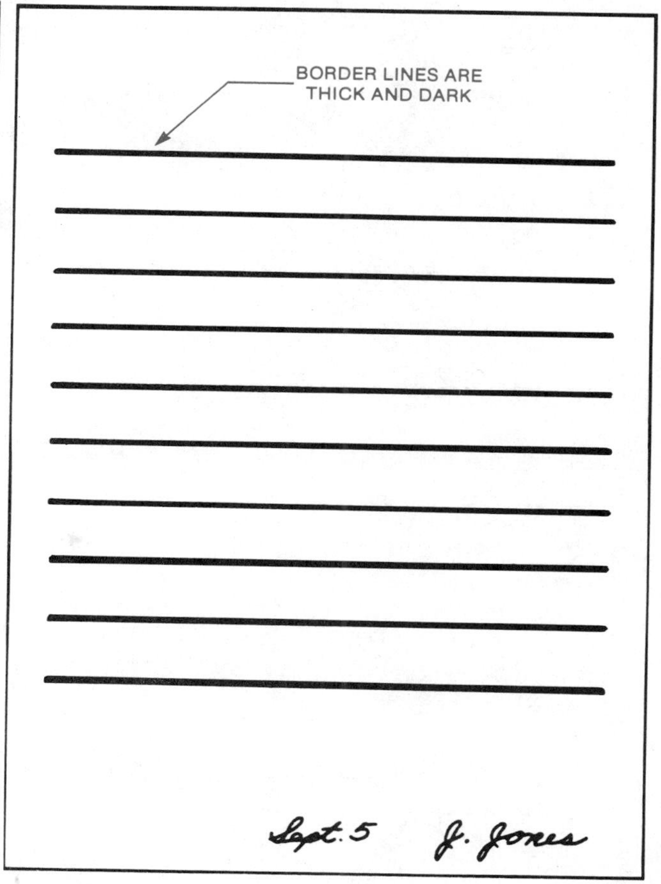

Figure 9-10
Step 2, *horizontal object lines.*

PRACTICE SHEET 9-B PRACTICING HORIZONTAL BORDER LINES

Step 1. Draw ten parallel construction lines horizontally (Figure 9-11). Space the lines to fill the entire sheet.

Step 2. Darken each construction line (Figure 9-12) to the width of a border line (Figure 9-3). Point your pencil whenever needed.

REPOINT PENCIL
OFTEN
Sept. 8
S. Smith

Figure 9-11
Step 1, *horizontal construction lines.*

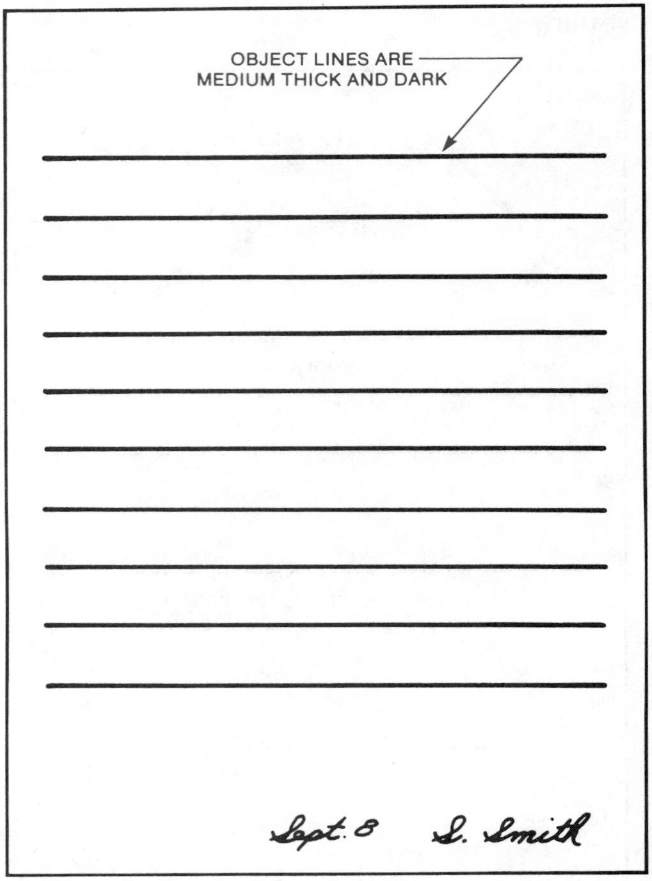
OBJECT LINES ARE
MEDIUM THICK AND DARK
Sept. 8
S. Smith

Figure 9-12
Step 2, *horizontal border lines.*

DRAWING VERTICAL LINES

Vertical lines are drawn with a T-square, a pencil, and either a 45° or 30°-60° triangle. Bring the T-square into position and hold it firmly against the drawing board edge. Then slide the hand holding the T-square over to hold the triangle along the top edge of the T-square (Figure 9-13). Slant the pencil away from the vertical edge of the triangle and in the direction it will be pulled (Figure 9-14). Pull upward and rotate the pencil slowly (Figure 9-15). Make sure that the T-square head is firmly against the drawing board edge and that the triangle under your hand is held firmly along the top of the T-square. This will keep each vertical line parallel to the rest.

To move the triangle to a new position, lift the triangle slightly and slide it along the T-square (Figure 9-16). Sliding the triangle without lifting it will smear the lines.

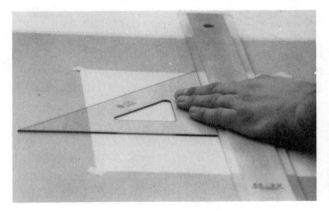

Figure 9-13
Align the triangle with the top edge of the T-square.

Figure 9-14
Slant the pencil away from the triangle and in the pulling direction.

Figure 9-15
Rotate the pencil while pulling.

Figure 9-16
Lift and slide the triangle along the T-square.

PRACTICE SHEET 9-C PRACTICING VERTICAL OBJECT LINES

Step 1. Draw ten vertical construction lines (Figure 9-17). Be sure the T-square and triangle are held firmly in position. Lift the triangle slightly when moving it.

Step 2. After pointing your pencil, darken each construction line to the width of an object line (Figure 9-18). Point your pencil when needed.

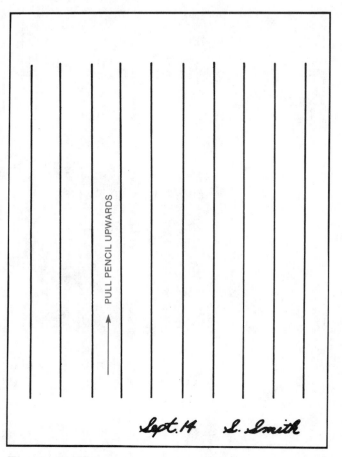

Figure 9-17
Step 1, *vertical construction lines.*

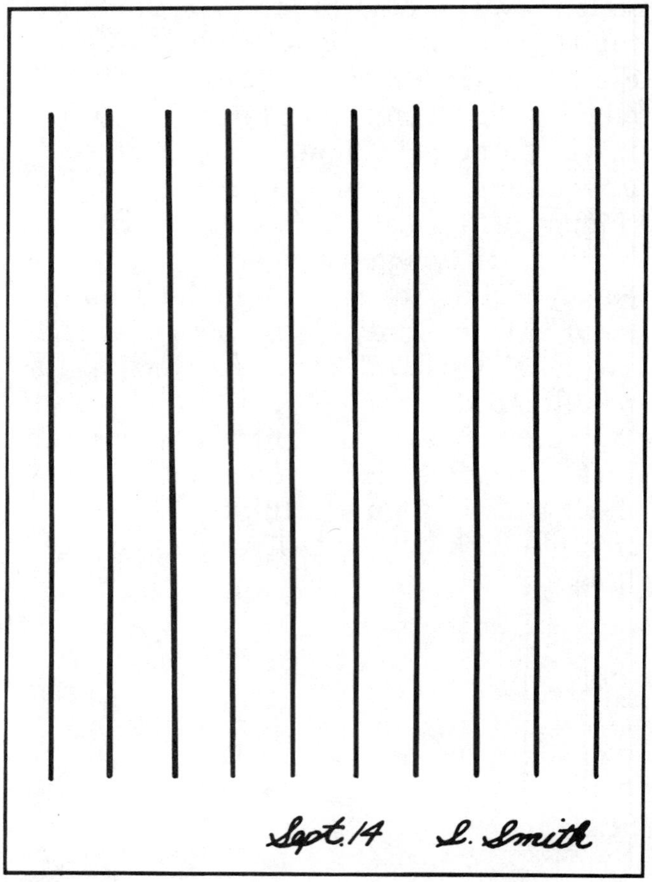

Figure 9-18
Step 2, *vertical object lines.*

PRACTICE SHEET 9-D PRACTICING VERTICAL BORDER LINES

Step 1. Prepare your sheet as in the previous exercise. Draw ten vertical construction lines (Figure 9-19).

Step 2. Darken each construction line to the width of a border line (Figure 9-20). Point your pencil as needed.

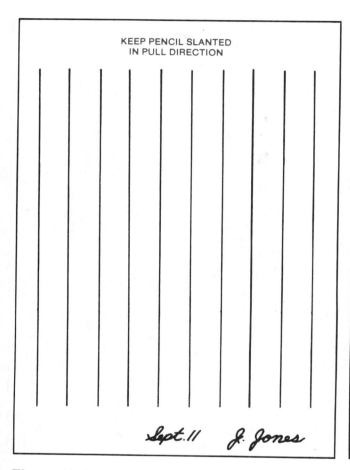

Figure 9-19
Step 1, *vertical construction lines.*

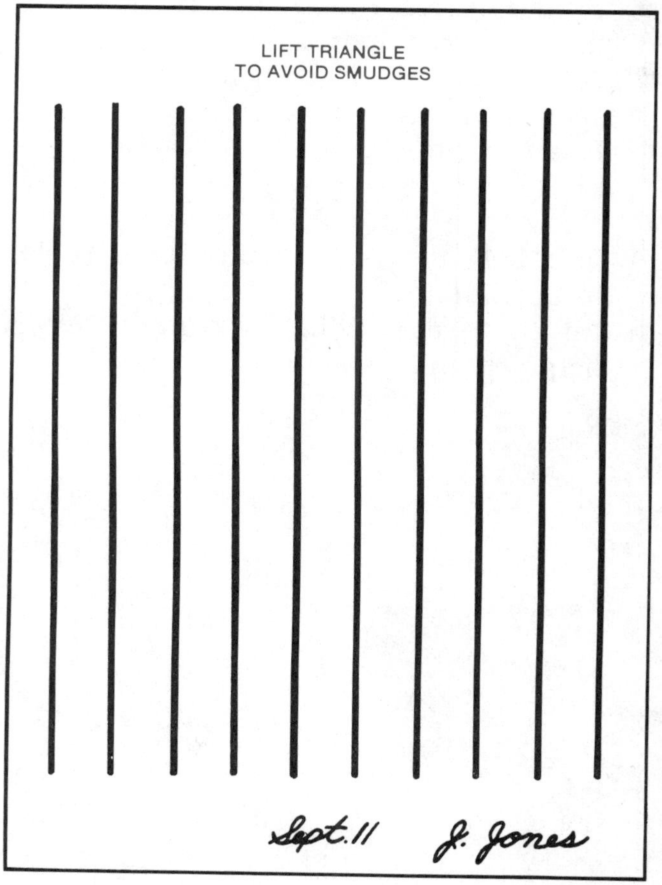

Figure 9-20
Step 2, *vertical border lines.*

- Horizontal and vertical lines should be uniform.
- Light construction lines should always be drawn first.
- Object lines are medium thick.
- Border lines are heavier than construction or object lines.
- A T-square should be held firmly against the drawing board edge.
- Pencil rotation and tilt are important in all line drawing.
- For vertical lines, it is important to hold both the T-square and the triangle in position.
- Lifting triangles while moving them helps eliminate smudges.

INCLINED LINES

This unit shows you how to draw inclined lines. You will discover:

- how to use a T-square, a pencil, and triangles to draw inclined lines,
- how to draw the inclined lines most common in mechanical drawing,
- how to draw accurate inclined lines.

KEY WORDS

Inclined: A line set at an angle from vertical or horizontal (Figure 10-1).
Angle: The shape made by two straight lines meeting at a point (Figure 10-1).
Degree: A unit of measurement for angles. The symbol for degree (°) is usually used on the drawing itself. Circles contain 360 degrees (360°) (Figure 10-1).

DRAWING INCLINED LINES

On mechanical drawings, certain *inclined* lines are used more often than others. Usually inclined lines at *angles* of 15, 30, 45, 60, 75, and 90 *degrees* are most common (Figure 10-2). With a T-square, a

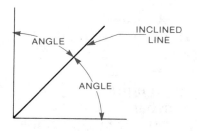

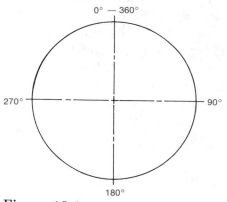

Figure 10-1
Top, *an inclined line and an angle;* bottom, *a circle contains 360 degrees.*

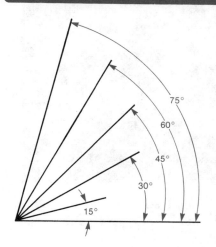

Figure 10-2
Inclined lines: 15 degrees, 30 degrees, 45 degrees, 60 degrees, and 75 degrees.

45 degree triangle, and a 30-60 degree triangle, you can draw any of the common inclined lines (Figure 10-3).

Drawing Lines 45 Degrees from the Horizontal. A 45 degree line can be inclined in any of four different directions (Figure 10-4). Lines 45 degrees are drawn with a 45 degree triangle, the upper edge of a T-square, and a pencil. Bring the T-square into position and hold it firmly against the drawing board edge. Then slide your hand from the T-square over to hold the triangle against the top edge of the T-square (Figure 10-5). Slant the pencil in the direction of pull and also away from the 45 degree edge of the triangle (Figure 10-6). Pull the pencil upward while rotating it slowly. Be sure to hold the T-square and triangle firmly under your hand. When you move the triangle, lift it slightly.

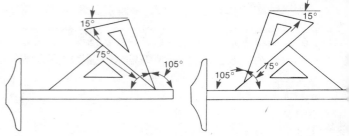

Figure 10-3
By using a 45 and a 30 degree triangle together, you can draw other angles.

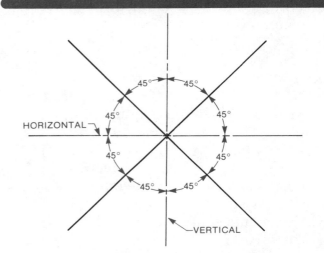

Figure 10-4   Lines inclined 45 degrees.

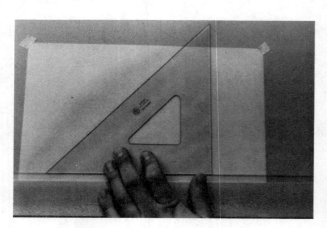

Figure 10-5   *Holding T-square and triangle.*

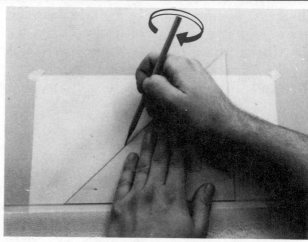

Figure 10-6   *Drawing an inclined line.*

Drawing Lines 30 Degrees from the Horizontal. Lines 30 degrees from the horizontal can be inclined in one of four directions (Figure 10-7). They are drawn with a 30-60 degree triangle, the upper edge of a T-square, and a pencil. Place the triangle along the upper edge of the T-square as shown in Figures 10-8 and 10-9. To draw lines 30 degrees from the horizontal, slant, pull, and rotate the pencil as you did in previous exercises.

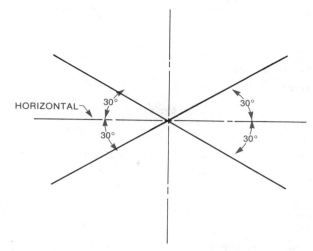

Figure 10-7 Lines inclined 30 degrees.

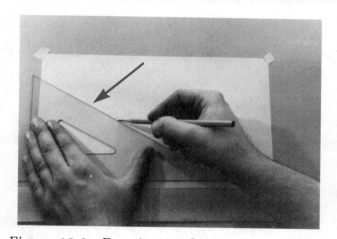

Figure 10-8 *Drawing 30 degree inclined lines.*

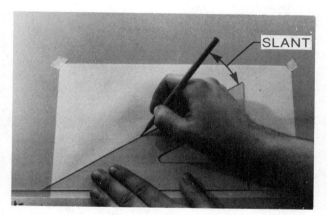

Figure 10-9 *Drawing inclined lines.*

Drawing Lines 60 Degrees from the Horizontal.   There are four directions from the horizontal that 60 degree lines can be drawn (Figure 10-10). To draw inclined lines 60 degrees from the horizontal, place a 30-60 degree triangle as shown in Figure 10-11. Proceed as in previous exercises.

Drawing Lines 30 Degrees from the Vertical.   Lines 30 degrees from the vertical are actually the same as lines angled 60 degrees from the horizontal (Figures 10-12, 10-13, 10-14, and 10-15). Thus, lines 30 degrees from the vertical are drawn as lines 60 degrees from the horizontal.

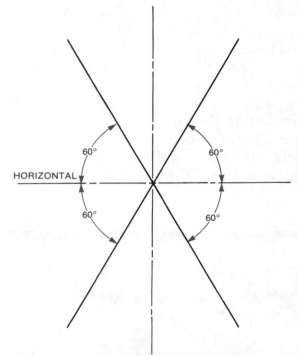

Figure 10-10   *Lines inclined 60 degrees.*

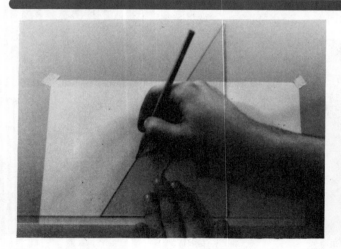
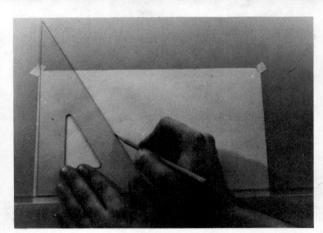

Figure 10-11   *Drawing 60 degree inclined lines.*

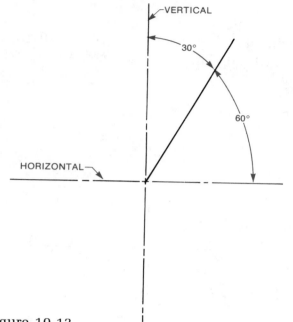

Figure 10-12
Line 30 degrees right from vertical and 60 degrees up from horizontal.

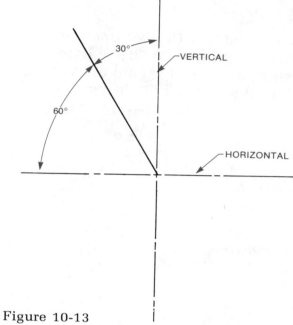

Figure 10-13
Line 30 degrees left from vertical and 60 degrees up from horizontal.

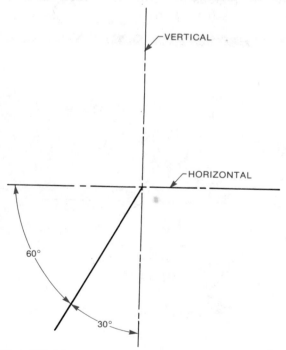

Figure 10-14
Line 30 degrees left from vertical and 60 degrees down from horizontal.

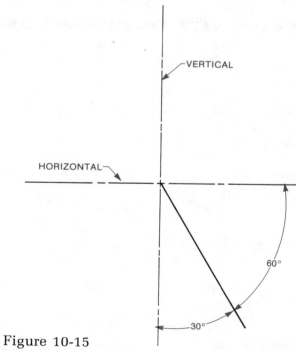

Figure 10-15
Line 30 degrees right from vertical and 60 degrees down from horizontal.

INCLINED LINES   69

Drawing Lines 60 Degrees from the Vertical. The four 60 degree lines drawn from the vertical are the same as the four lines angled 30 degrees from the horizontal (Figures 10-16, 10-17, 10-18, and 10-19). They are drawn the same way.

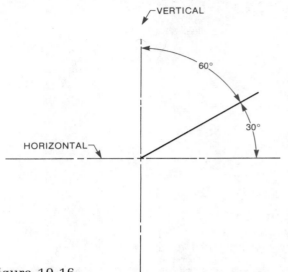

Figure 10-16
Line 30 degrees up from horizontal and 60 degrees right from vertical.

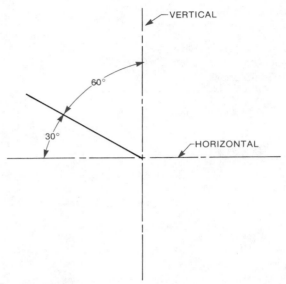

Figure 10-17
Line 30 degrees up from horizontal and 60 degrees left from vertical.

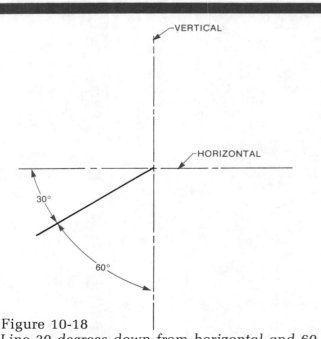

Figure 10-18
Line 30 degrees down from horizontal and 60 degrees left from vertical.

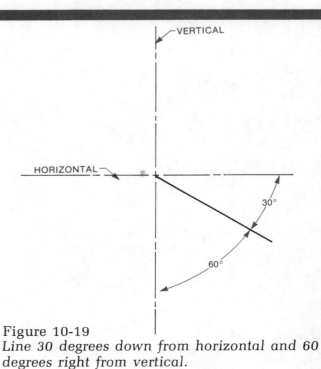

Figure 10-19
Line 30 degrees down from horizontal and 60 degrees right from vertical.

PRACTICE SHEET 10-A PRACTICING 45 DEGREE, 30 DEGREE, AND 60 DEGREE INCLINED LINES

Step 1. Divide a sheet of paper into four sections (two across and two down). This can be done by folding the paper lengthwise and then crosswise.

Step 2. Draw light construction lines using the angles indicated in Figure 10-20.

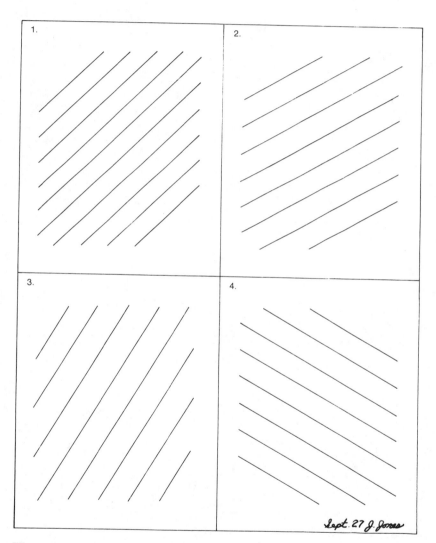

Figure 10-20   Step 2, *light construction lines.*

Step 3.   Darken each construction line to the width of an object line (Figure 10-21).

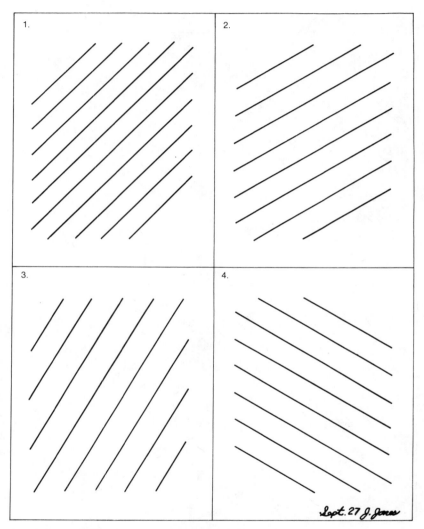

Figure 10-21   Step 3, *object lines.*

Drawing 15 Degree and 75 Degree Lines. To draw lines inclined at 15 and 75 degrees (Figure 10-22) use both the 45 degree and the 30-60 degree triangles. By arranging the triangles on your T-square, you can draw either a 15 degree or a 75 degree angle from horizontal or vertical (Figures 10-23 and 10-24). When drawing lines using both triangles, be sure the triangles are firmly

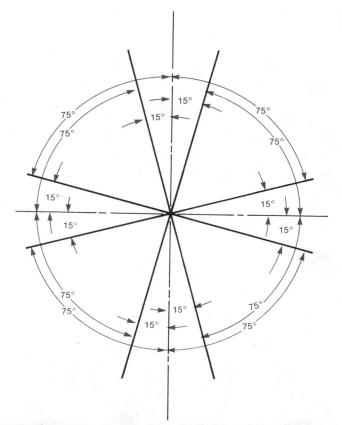

Figure 10-22
Inclined lines 15 degrees and 75 degrees.

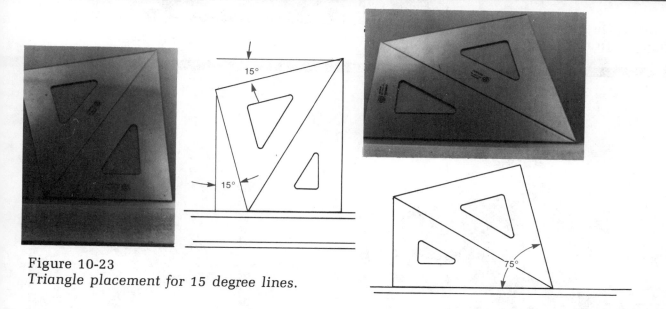

Figure 10-23
Triangle placement for 15 degree lines.

Figure 10-24
Triangle placement for 75 degree lines.

placed against each other (Figure 10-25) and the top of the T-square (Figure 10-26).

THE ADJUSTABLE TRIANGLE

By using an adjustable triangle (Figure 10-27), you can draw any inclined line by opening the triangle to the desired angle (Figure 10-28).

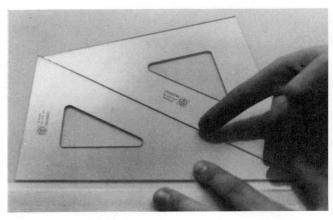

Figure 10-25
Triangles firmly against each other.

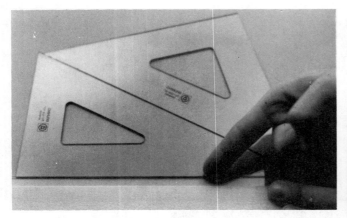

Figure 10-26
Triangles firmly on top of T-square.

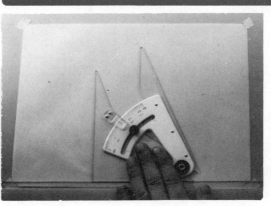

Figure 10-27   *An adjustable triangle.*

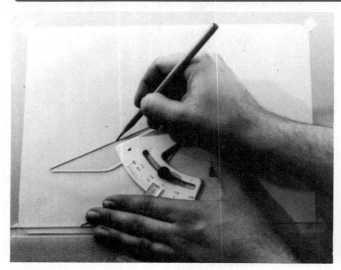

Figure 10-28   *Using the adjustable triangle.*

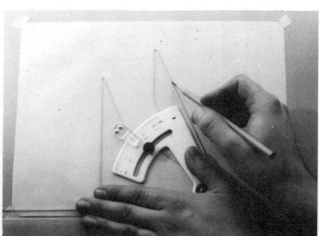

Step 1.   Locate the exact center of a sheet of paper by drawing two light construction lines connecting opposite corners of the sheet (Figures 10-29, 10-30, and 10-31).

Step 2.   Draw 24 construction lines from the center of your sheet, each 15 degrees from the next (Figure 10-32).

Step 3.   Darken each construction line to the width of an object line (Figure 10-33).

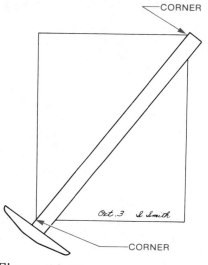

Figure 10-29
Step 1, using *T-square to connect corners.*

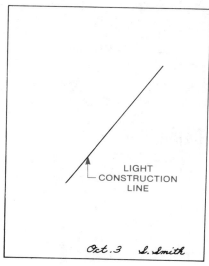

Figure 10-30
Step 2, *a light construction line.*

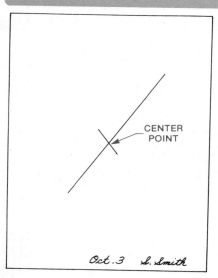

Figure 10-31
Step 3, *a second construction line crossing the first.*

Figure 10-32
Step 4, *construction lines 15 degrees apart.*

Figure 10-33
Step 5, *construction lines darkened to object lines.*

- Lines inclined at 15 degrees, 30 degrees, 45 degrees, 60 degrees, 75 degrees, and 90 degrees are most common on mechanical drawings.
- Inclined lines are drawn with the triangles held firmly along the top of the T-square.
- Lines 30 degrees from the horizontal are the same as lines 60 degrees from the vertical.
- Lines 60 degrees from the horizontal are the same as lines 30 degrees from the vertical.
- Lines 15 and 75 degrees are drawn with both a 45 and 30-60 degree triangle.
- An adjustable triangle can be adjusted for drawing any inclined line.

THE SCALE

This unit studies the *full* scale (1:1). You will discover:
● the basic use of the scale,
● how to measure distances accurately.

KEY WORDS
Proportion: Relative size of things, one compared to another.
Millimetre: A unit of length measurement used on mechanical drawings (Figure 11-1).

MEASUREMENTS
To draw accurate drawings, you must measure accurately. Drafters' success in drawing is based largely upon their ability to measure distances. The measuring tool, called a scale, has several edges (Figure 11-2). Each edge is called a *proportion* scale.

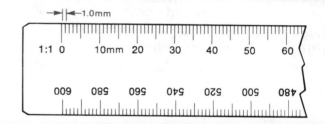

Figure 11-1
A millimetre on a scale. (AM Bruning)

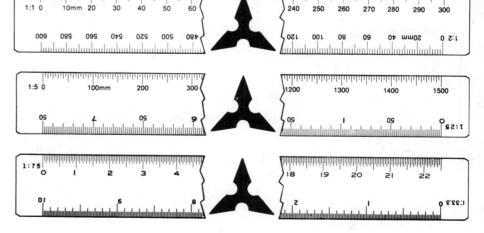

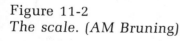

Figure 11-2
The scale. (AM Bruning)

Each proportion scale is designed to help the drafter reduce or enlarge the drawing of an object in proportion to its actual size. Figure 11-3 shows a 1:2 (half size) proportion, that is, the smaller figure is reduced proportionately to half size.

THE FULL SCALE

The full scale is marked 1:1 (Figure 11-4). A full scale does not reduce or enlarge the drawing of the object. Thus, all measurements made using the full scale are the actual size of the object, or in a proportion of one to one.

Using the Full Scale. All measurements begin at the first mark in from the edge of the scale (Figure 11-5). This mark is the starting point, or zero.

As you move toward the middle of the scale, each mark you pass represents one *millimetre* from zero (Figure 11-6). At the fifth mark from zero, the marking is made

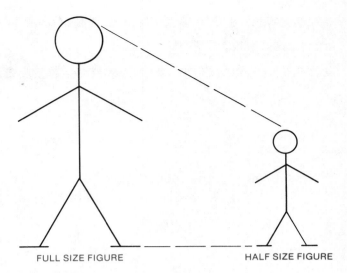

FULL SIZE FIGURE   HALF SIZE FIGURE

Figure 11-3
The small and large stick figures are in proportion to each other.

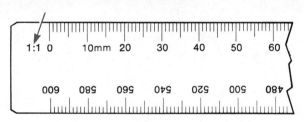

Figure 11-4
Full scale markings 1:1. (AM Bruning)

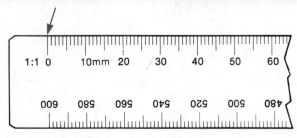

Figure 11-5
The zero mark. (AM Bruning)

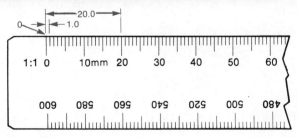

Figure 11-6
Distances from zero. (AM Bruning)

slightly longer than the other millimetre markings (Figure 11-7). This longer marking enables you to identify units of five millimetres easily and count out long measurements at the rate of five millimetres (Figure 11-8).

The tenth millimetre mark from zero is made longer than both the millimetre marking and the five millimetre marking (Figure 11-9). This marking enables the drafter to easily identify units of ten millimetres and also to count out longer distances by units of ten millimetres (Figure 11-10).

On many drawings, it is necessary to measure a distance to a half millimetre (0.5 mm). Since only full millimetre distances from zero have markings on the scale, the drafter must estimate the midpoint between markings for the half millimetre (Figure 11-11).

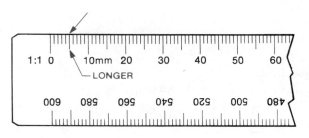

Figure 11-7
The fifth mark from zero. (AM Bruning)

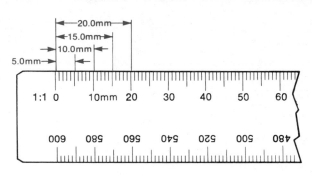

Figure 11-8
Units of five millimetres. (AM Bruning)

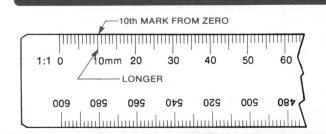

Figure 11-9
The tenth mark from zero. (AM Bruning)

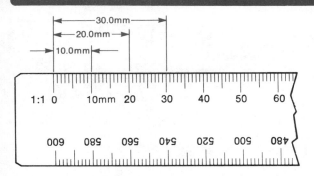

Figure 11-10
Units of ten millimetres. (AM Bruning)

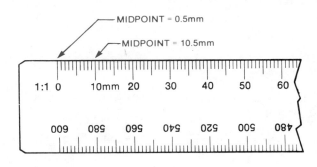

Figure 11-11
*Midpoints between millimetre marks.
(AM Bruning)*

Using your scale, measure each of the line lengths here. Be sure to put the zero or beginning mark of your scale at one end of the line. The example shown is *30.0 mm.* Write your answers on a separate piece of paper.

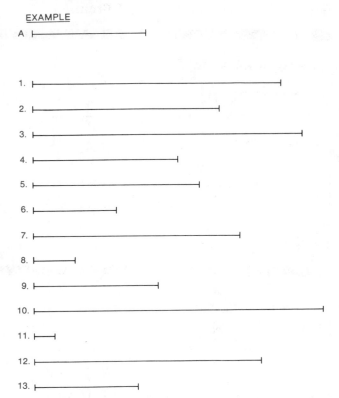

MARKING OFF DISTANCES

You should use a very sharp pencil (Figure 11-12) to mark off distances on a drawing. Place the zero mark of your scale exactly at the beginning point (Figure 11-13). "Point" a light dot at that spot. Measure the distance and "point" the end point (Figure 11-14). Connect the two points with a light construction line and then darken the line.

Figure 11-12
A sharp pencil for marking off distances.

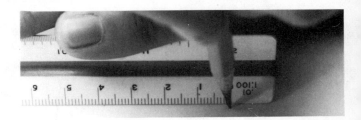

Figure 11-13
Zero at the beginning of the distance to be measured.

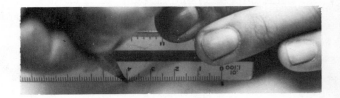

Figure 11-14 *Pointing the distance desired.*

On a separate sheet, measure and draw ten horizontal lines exactly as shown here.

Step 1. Begin all measurements at zero.

Step 2. Draw each line as a light construction line.

Step 3. Check each measurement before darkening your lines.

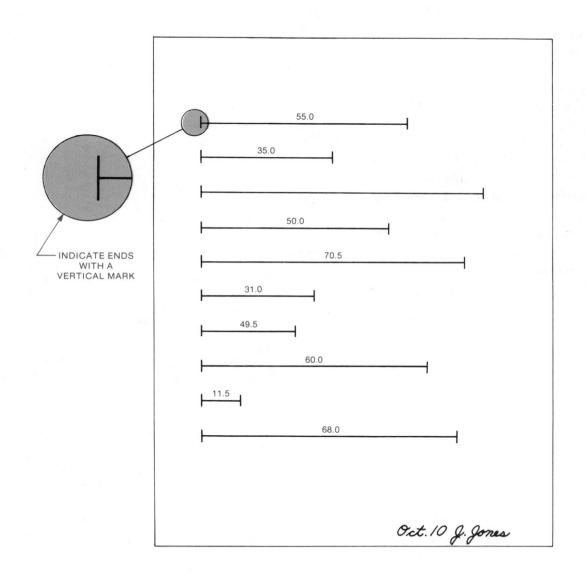

INDICATE ENDS
WITH A
VERTICAL MARK

55.0

35.0

50.0

70.5

31.0

49.5

60.0

11.5

68.0

Oct. 10 J. Jones

- In drafting, a scale is used for measuring.
- Each measuring edge of the scale is called a proportion scale.
- Each proportion scale helps the drafter reduce or enlarge the drawing of an object in relation to the object's actual size.
- The full scale is used to make actual size measurements.
- Each small marking on the full scale represents one millimetre in distance from the zero marking.
- Every fifth millimetre marking is larger than a single millimetre marking.
- Every tenth millimetre marking is larger than all other markings.
- Half millimetre distances are midway between millimetre markings.
- A very sharp pencil should be used for marking-off distances.

LETTERING

This unit shows you how to letter on drawings. You will discover:
- the forms of letters and numbers,
- how to make even, uniform letters and numbers,
- how to use equipment to aid in lettering.

KEY WORDS
Diagonal: A slanting line, from one corner to the opposite corner (Figure 12-1).
Oval: Egg-shaped (Figure 12-2).
Unit: A fixed distance used to proportion letters and numbers (Figure 12-3).

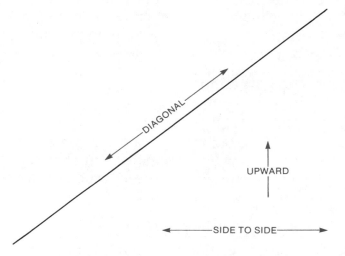

Figure 12-1 *A diagonal line.*

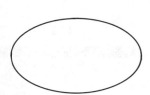

Figure 12-2 *An oval shape.*

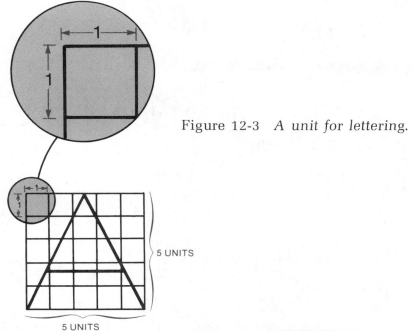

Figure 12-3 *A unit for lettering.*

LETTERS AND NUMBERS IN DRAFTING

A complete mechanical drawing shows not only the object; the drawing also gives information that cannot be shown by lines alone. This information is given in lettered notes and numbered dimensions (Figure 12-4). Since the lettering is an important part of the drawing, it *must* look good and be read easily. To look good, lettering should be uniform in height and width and of even line weight.

Like any skill, lettering takes practice to develop. Before you can practice and develop your lettering skill, you should know the correct proportions and forms of letters.

THE GOTHIC LETTERS

The lettering style most widely used in mechanical drawing is the vertical Gothic style letter (Figure 12-5). Gothic letters are made up of combinations of single lines. These lines may be horizontal, vertical, *diagonal*, curved, or *oval* in form. Only capital (uppercase) letters are used.

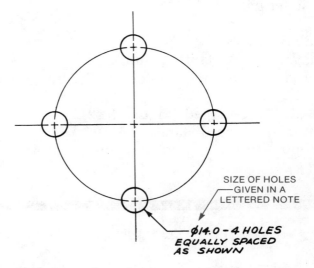

SIZE OF HOLES
GIVEN IN A
LETTERED NOTE

Ø14.0 – 4 HOLES
EQUALLY SPACED
AS SHOWN

Figure 12-4 *A drawing with a lettered note.*

ABCDEFGHIJKLMNOPQRSTUVWXYZ
1234567890

Figure 12-5 *Gothic style letters and numbers.*

GUIDELINES

To control the size and proportion of letters and numbers, you should draw guidelines. Use horizontal guidelines to control letter height (Figure 12-6). Random vertical guidelines control spacing and vertical straightness (Figure 12-7). All guidelines are drawn with a sharp 3H pencil and a T-square and lettering guide (Figure 12-8).

PENCILS FOR LETTERING

An H pencil should be used for all lettering. Round the point slightly (Figure 12-9) for a thick-bodied letter.

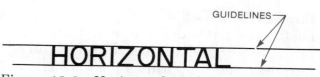

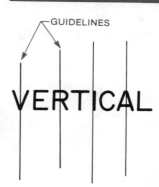

Figure 12-6   *Horizontal guidelines.*

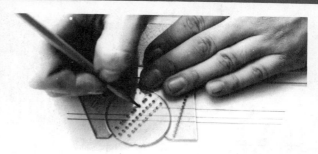

Figure 12-7   *Vertical guidelines.*

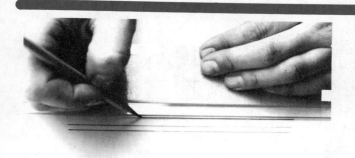

Figure 12-8   Left, *using a T-square; right, a lettering guide for drawing guidelines.*

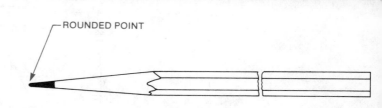

Figure 12-9
A rounded point on a lettering pencil.

FORMING LETTERS

Each letter has a definite height in proportion to a definite width. For our purposes, the proportions for the height and width of letters will be expressed in imaginary *units*.

Figure 12-10 shows the layout of straight line letters. Follow the numbered sequence in drawing the lines. The arrows show how the line is made. For example, the letter *L* has one downstroke (1) then one side stroke (2). Follow arrows. The *L* is 5 units high and 4 units wide.

Figure 12-10   *Straight line letters.*

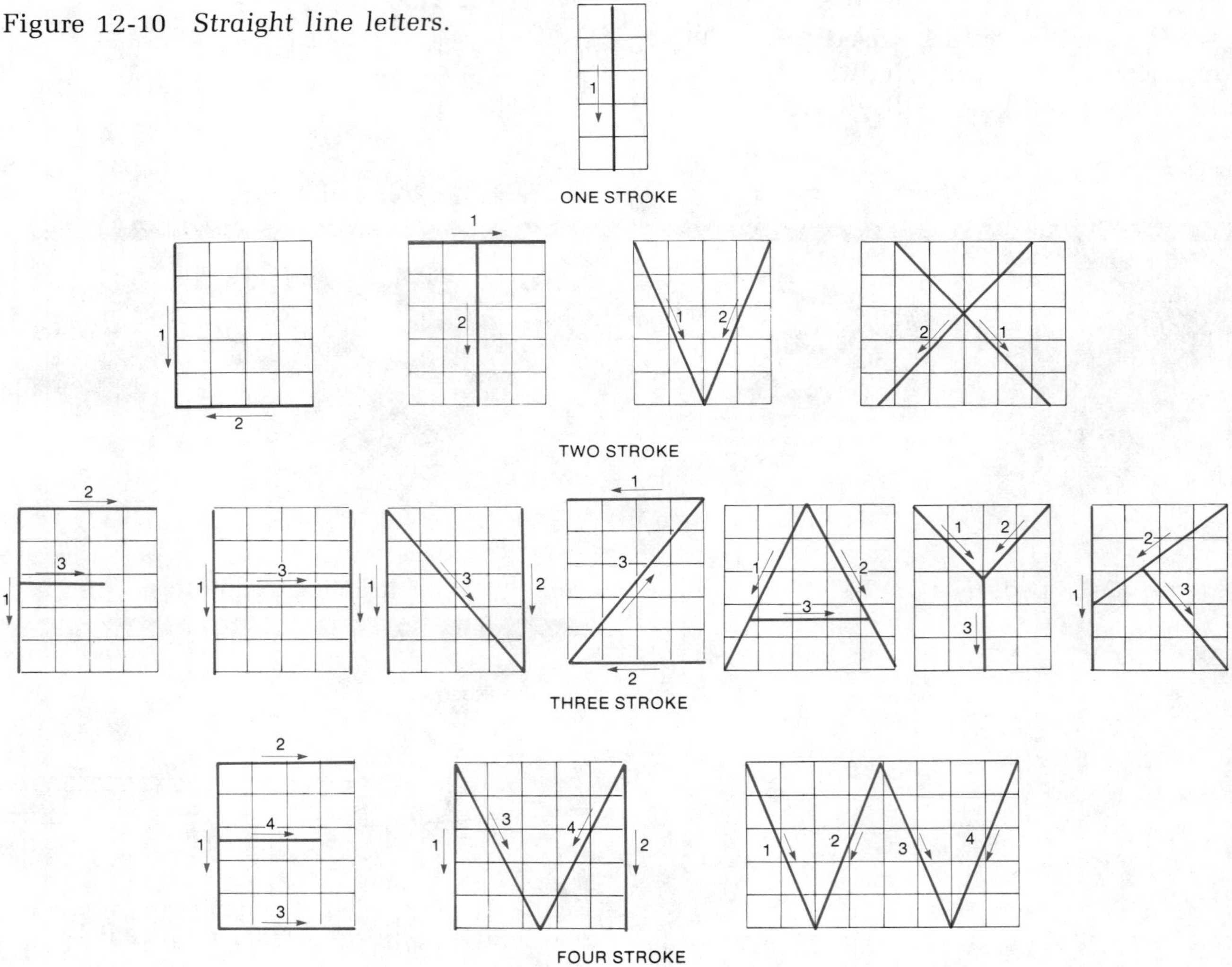

ONE STROKE

TWO STROKE

THREE STROKE

FOUR STROKE

Properly mount and prepare an A4 sheet with guidelines as shown. Keep the guidelines very light. Remember your pencil point should be slightly rounded. Repeat the letters to fill the width of the page.

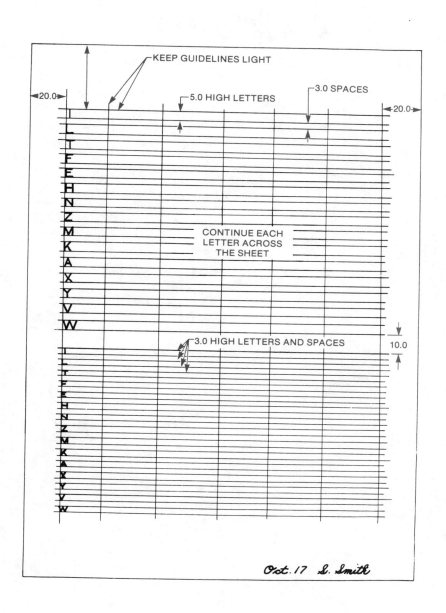

Curved letters are also formed in sequence. Figure 12-11 shows the drawing procedure.

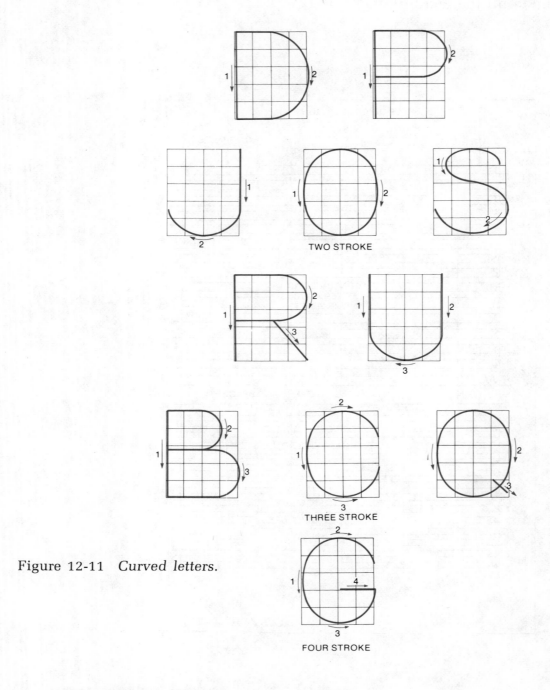

Figure 12-11 *Curved letters.*

On a properly mounted and prepared A4
sheet, practice curved letters as shown.

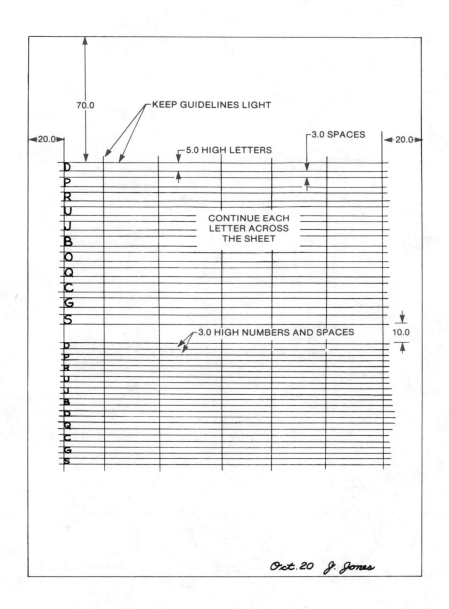

Numbers, like letters, are drawn in sequence. Figure 12-12 shows the drawing procedure.

ONE STROKE

Figure 12-12   *Numbers.*

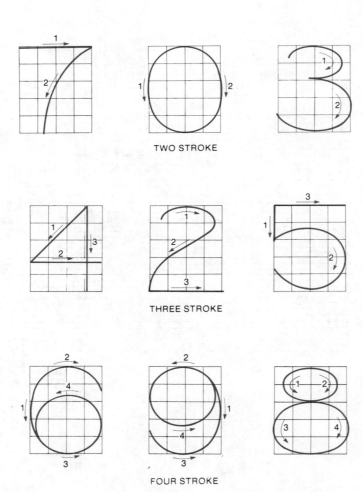

TWO STROKE

THREE STROKE

FOUR STROKE

On a properly mounted and prepared A4
sheet, letter the numbers as shown.

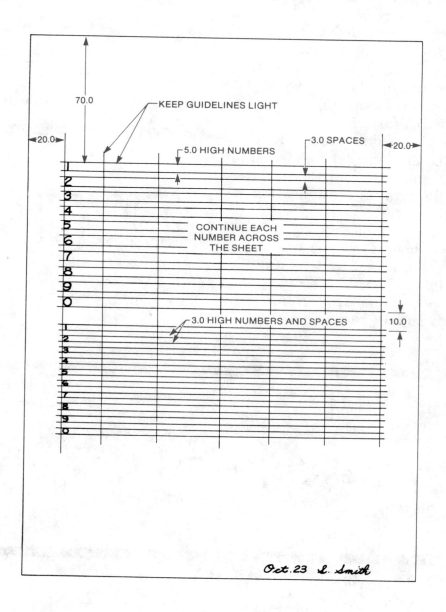

SPACING

Besides forming letters correctly, you must learn correct spacing so the lettering will be neat and easy to read. Depending upon the combinations of letters, different amounts of space are needed so the lettering will *appear* uniform. You will need to know how much space to allow between letters and numbers, between words, and between sentences.

Spacing Within Words. Between two letters with vertical sides, such as *H, I, M, N, T,* and *U,* and also letters *J* and *Z,* use a two unit space (Figure 12-13).

Spacing to the right of the following letters should be one-half unit (Figure 12-14): *B, D, E, F, K, L, P,* and *R.*

Spacing on either side of the following letters should be one-half unit (Figure 12-15): *A, C, G, O, Q, S, V, W, X,* and *Y.*

Spacing Between Words. Spacing between words should be four units (Figure 12-16).

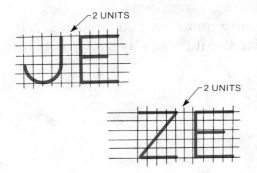

Figure 12-13 *Spacing between letters J and Z.*

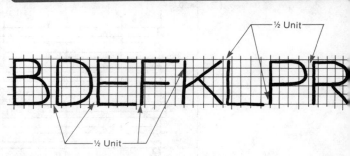

Figure 12-14
Spacing between letters B, D, E, F, K, L, P, and R.

Figure 12-15
Spacing between letters A, C, G, O, Q, V, W, and X.

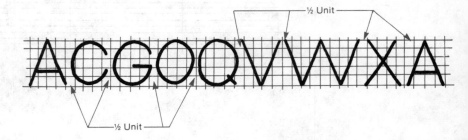

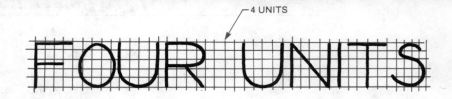

Figure 12-16
Spacing between words.

Spacing Between Sentences. Spacing between sentences should be eight units (Figure 12-17).

Spacing Between Numbers. Spacing on either side of the following numbers should be one-half space (Figure 12-18): 0, 6, 7, and 9.

Spacing of the number 4 should be one-half space on the left and two units on the right.

Spacing on either side of the following numbers should be two units (Figure 12-19): 1, 2, 3, 5, and 8.

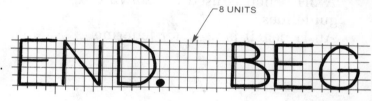

Figure 12-17   *Spacing between sentences.*

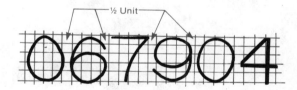

Figure 12-18
Spacing between numbers 0, 4, 6, 7, and 9.

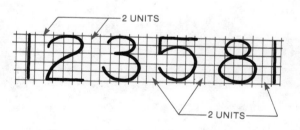

Figure 12-19
Spacing between numbers 1, 2, 3, 5, and 8.

- Lettering should be uniform in height, width, and thickness of line.
- Gothic capital letters are used on mechanical drawings.
- Light guidelines are used to control height, spacing, and vertical straightness.
- Each letter and number has a definite width in proportion to height.
- Spacing between letters and numbers differs.
- Spacing between words is generally four units.
- Spacing between sentences should be eight units.
- A 3H pencil is used for drawing guidelines.
- An H pencil is used for lettering.

FLAT LAYOUTS

This unit shows you how to make simple drawings. You will learn:
- how to draw single view drawings,
- how to use the drafting tools accurately in the preparation of drawings,
- how to prepare a simple drawing.

KEY WORDS

Thickness: The third dimension of an object; it is not the height or length (Figure 13-1).

Working Space: The portion of a drawing sheet that is inside the borders (Figure 13-2).

Title: The name of the drawing. The title of a drawing is lettered in the lower right-hand corner of the working space above the title block.

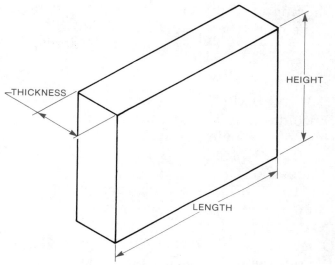

Figure 13-1   *Thickness.*

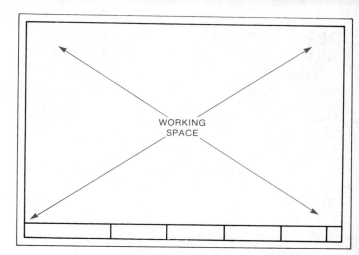

Figure 13-2   *Working space.*

FLAT LAYOUTS

The *thickness* of a thin object is often difficult to represent on a drawing (Figure 13-3). Since a slight thickness (12.0 mm or less) would offer little or no value, it is not shown on the drawing. Drawings that do not show the thickness of the objects are called flat layouts. Items 12.0 mm thick or less are usually drawn as flat layouts.

THE ALPHABET OF LINES

Since drafting is a graphic language, the lines of a drawing become the "alphabet." The alphabet of lines for flat layouts includes:

Construction Lines. Construction lines, as described in earlier units, are drawn very thin and light (Figure 13-4).

Object Lines. The lines that represent the outline of the object are darkened construction lines. Object lines are drawn dark and sharp and of medium thickness (Figure 13-5).

Border Lines. The lines that frame the drawing are darkened construction lines. Border lines are drawn thick and dark (Figure 13-5). They are heavier than object lines.

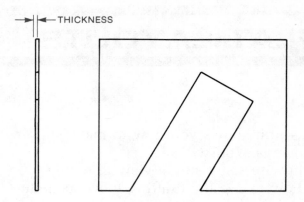

Figure 13-3
Little information is given by showing the thickness of a thin object.

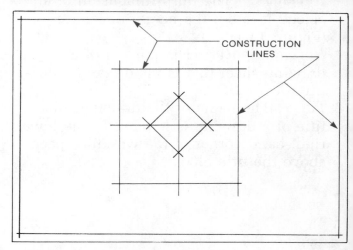

Figure 13-4   *Construction lines.*

Center Lines. Center lines are used to show the exact center of objects (Figure 13-5) and the center of portions of objects (Figure 13-6). Center lines are drawn thin, dark, and sharp (Figure 13-7).

PREPARATION BEFORE DRAWING FLAT LAYOUTS

All of the preparation steps discussed in the earlier units are now required to prepare a complete flat layout drawing. The following steps will serve as a review for preparation before drawing:

Step 1. Select the proper sheet size for the item to be drawn. Note: the sheet size required for each drawing in this text is given with the instructions for drawing.

Step 2. If the drawing sheet does not have preprinted borders and title strip, draw them.

Step 3. Enter title strip information. Be sure to use light guidelines for vertical and horizontal control of lettering.

Step 4. The drawing sheet should then be mounted and conditioned.

Step 5. Sharpen, point, and clean pencils that will be used for line work and lettering.

After you have completed the preparation, begin the drawing.

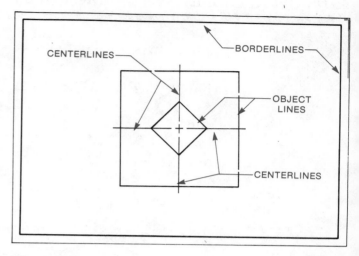

Figure 13-5
Object lines, border lines, and center lines.

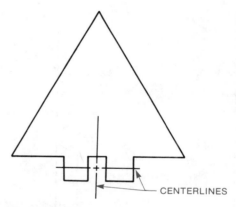

Figure 13-6
Center lines for a portion of an object.

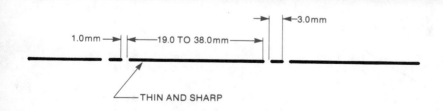

Figure 13-7   *Center line sizes.*

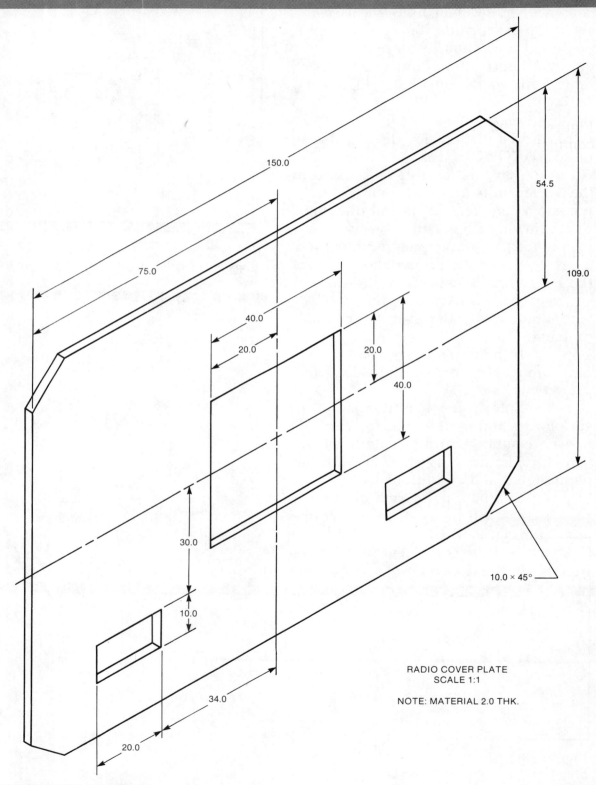

150.0

75.0

54.5

109.0

40.0

20.0

20.0

40.0

30.0

10.0

34.0

20.0

10.0 × 45°

RADIO COVER PLATE
SCALE 1:1

NOTE: MATERIAL 2.0 THK.

Figure 13-8 shows the steps needed to complete the drawing of the Radio Cover Plate. Use an A4 sheet.

Step 1. Draw light diagonal lines connecting the corners of the *working space* to locate the center.

Step 2. Draw horizontal and vertical center lines through the point where the diagonal lines cross. Draw center lines as shown earlier in this Unit. Be sure to have the short dashes cross at the center.

Step 3. Using the full scale, measure and mark off the distances for each edge of the object.

Step 4. Draw all horizontal and vertical construction lines.

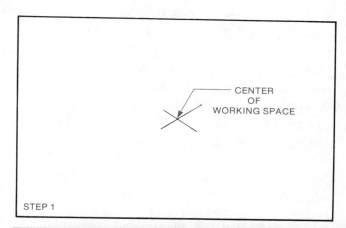

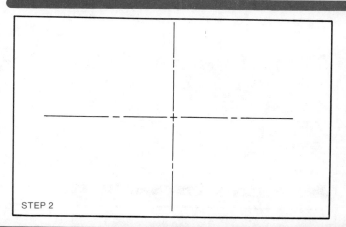

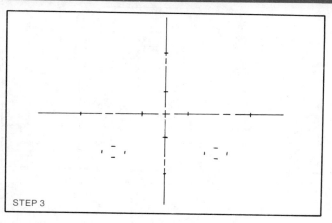

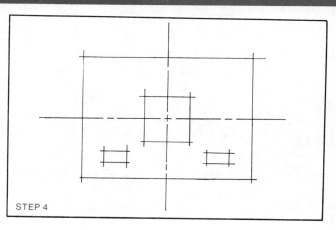

Figure 13-8
Steps to complete the drawing of the radio cover plate.

Step 5. Draw all inclined construction lines.

Step 6. Erase excess line work. Use your erasing shield where needed.

Step 7. Darken all line work.

Step 8. Enter *title* and scale information near the lower right-hand corner of the working space. Use light horizontal guidelines to control the lettering height. Lettering 5.0 mm high is used for drawing titles. Allow a 3.0 mm space below the title. Use 3.0 mm high letters for the drawing scale notation. Since the scale on this drawing is full scale, the notation should read: "Scale 1:1". Use light vertical guidelines for control.

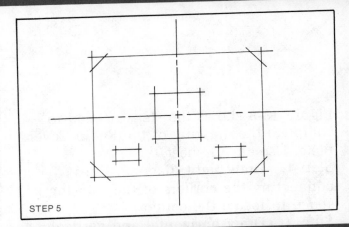

STEP 5

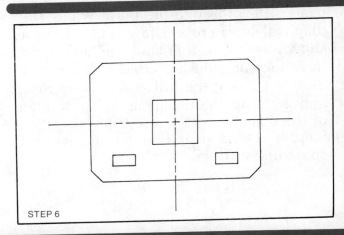

STEP 6

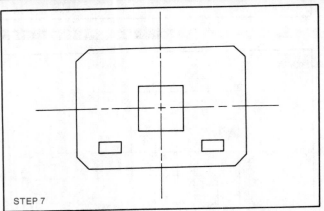

STEP 7

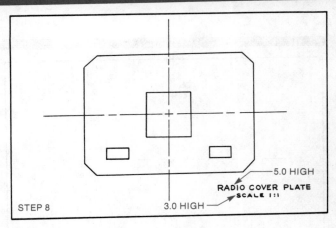

STEP 8

5.0 HIGH

RADIO COVER PLATE
SCALE 1:1

3.0 HIGH

Figure 13-8 (Continued)

Figure 13-9 shows steps needed to
complete the drawing of the Sheet Metal
Gage. Use an A4 sheet.

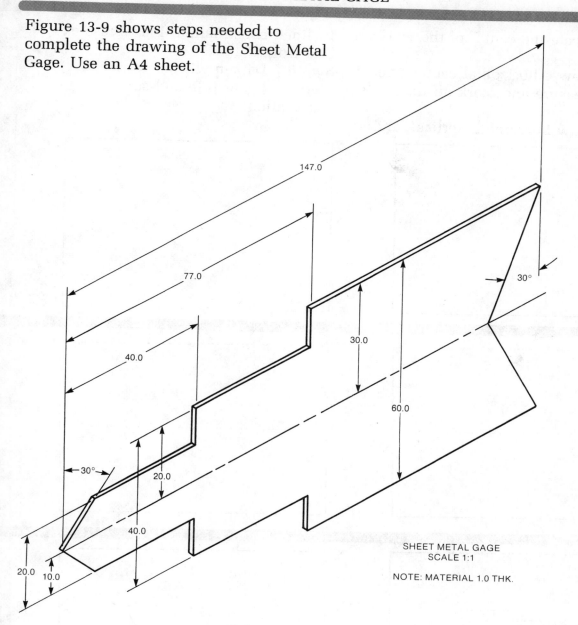

SHEET METAL GAGE
SCALE 1:1

NOTE: MATERIAL 1.0 THK.

Step 1.   Locate the center of the working space.

Step 2.   Draw a horizontal center line.

Step 3.   Measure and mark off distances on the full scale.

Step 4.   Draw horizontal, vertical, and inclined construction lines. Erase excess line work.

Step 5.   Darken all line work.

Step 6.   Letter title and scale information.

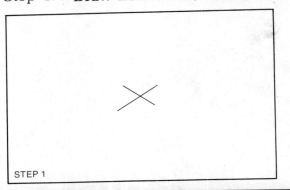

STEP 1

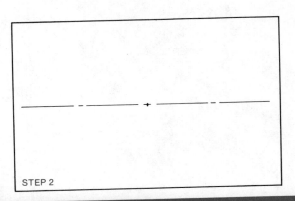

STEP 2

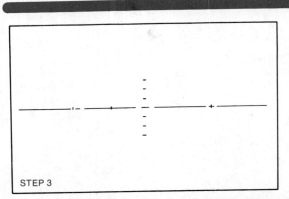

STEP 3

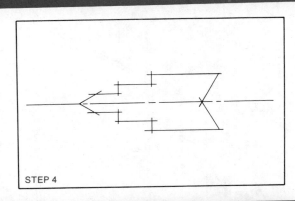

STEP 4

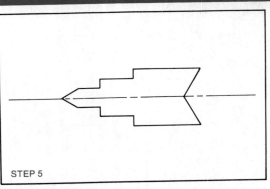

STEP 5

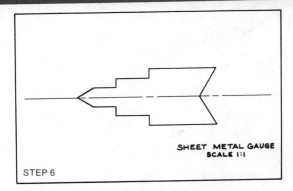

SHEET METAL GAUGE
SCALE 1:1

STEP 6

Figure 13-9
Steps needed to complete the drawing of the sheet metal gage.

DRAWING PROBLEM 13-3 SPACER PLATE

Use an A4 sheet. Locate and lay out the Spacer Plate in the center of the working space. Follow the construction steps as outlined for Drawing Problems 13-1 and 13-2.

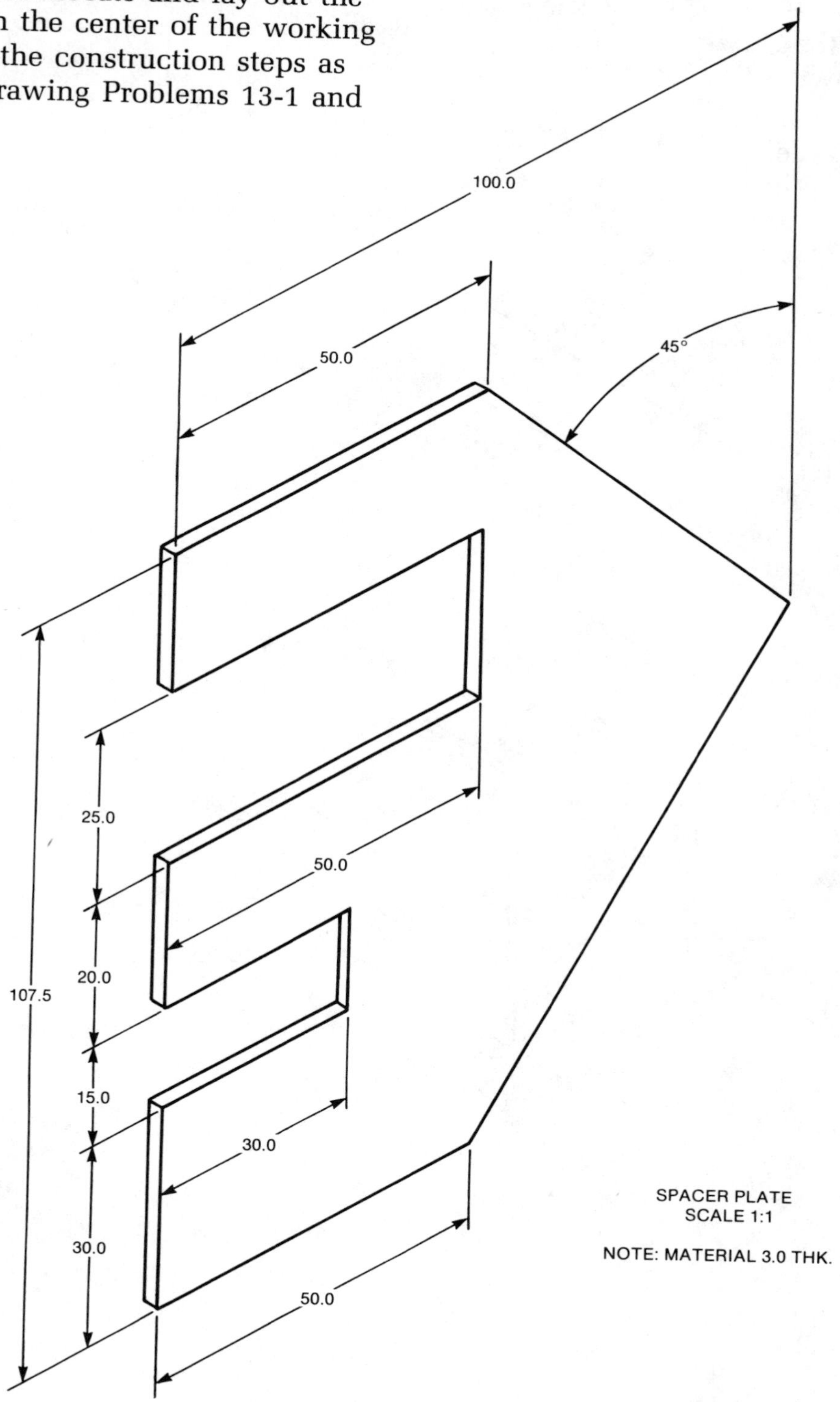

100.0
50.0
45°
25.0
50.0
20.0
107.5
15.0
30.0
30.0
50.0
SPACER PLATE
SCALE 1:1
NOTE: MATERIAL 3.0 THK.

Use an A4 sheet. Prepare a flat layout drawing for the Guide Plate. Review the steps for Drawing Problems 13-1, 13-2, and 13-3 before starting.

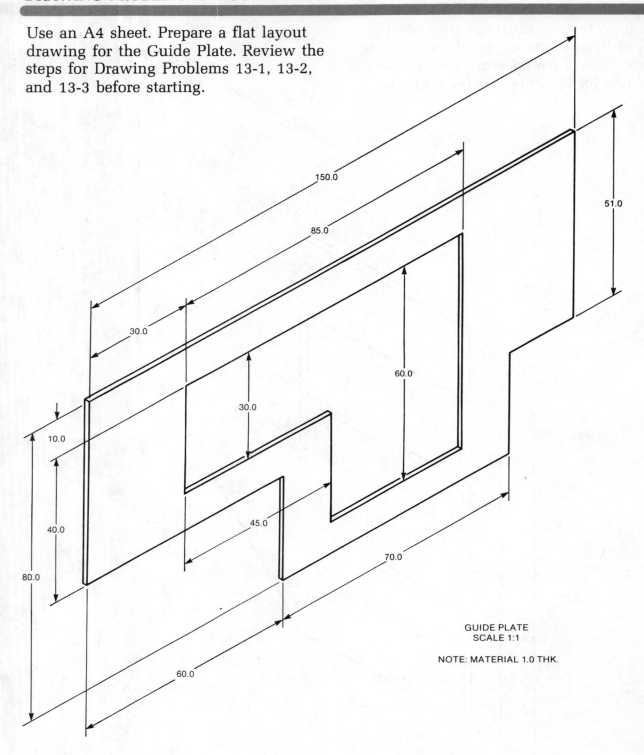

150.0

85.0

51.0

30.0

60.0

30.0

10.0

40.0

45.0

70.0

80.0

60.0

GUIDE PLATE
SCALE 1:1

NOTE: MATERIAL 1.0 THK.

DRAWING PROBLEM 13-5 30-60° TRIANGLE

On an A4 sheet prepare a drawing of the
30-60° Triangle.

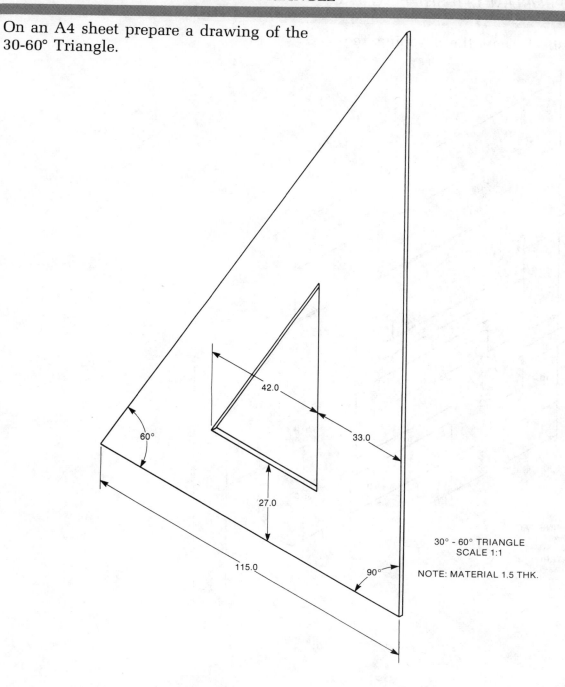

60°

42.0

33.0

27.0

115.0

90°

30° - 60° TRIANGLE
SCALE 1:1

NOTE: MATERIAL 1.5 THK.

Use an A4 sheet. Draw the Shipping Tag.

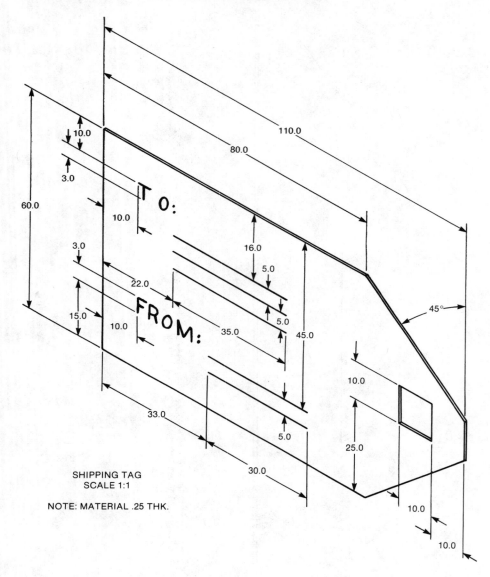

TO:

FROM:

110.0

80.0

10.0

3.0

60.0

10.0

3.0

22.0

15.0

10.0

16.0

5.0

5.0

35.0

45.0

45°

10.0

25.0

33.0

5.0

30.0

10.0

10.0

SHIPPING TAG
SCALE 1:1

NOTE: MATERIAL .25 THK.

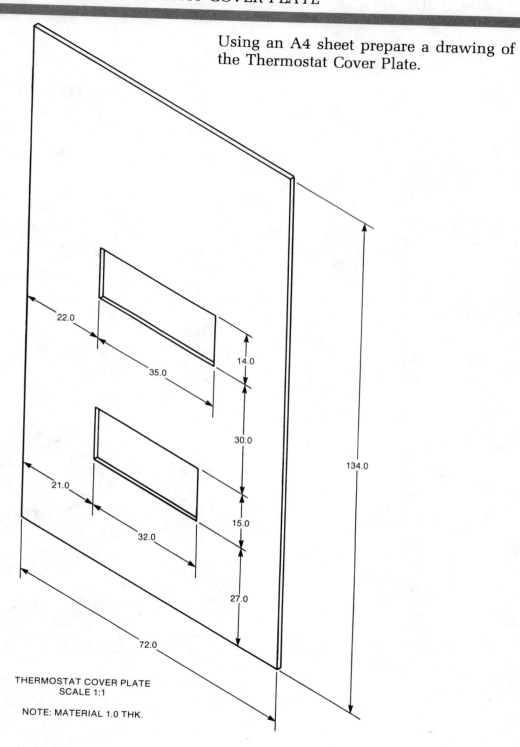

Using an A4 sheet prepare a drawing of the Thermostat Cover Plate.

22.0

35.0

14.0

30.0

134.0

21.0

32.0

15.0

27.0

72.0

THERMOSTAT COVER PLATE
SCALE 1:1

NOTE: MATERIAL 1.0 THK.

Draw the Key Blank on an A4 sheet.

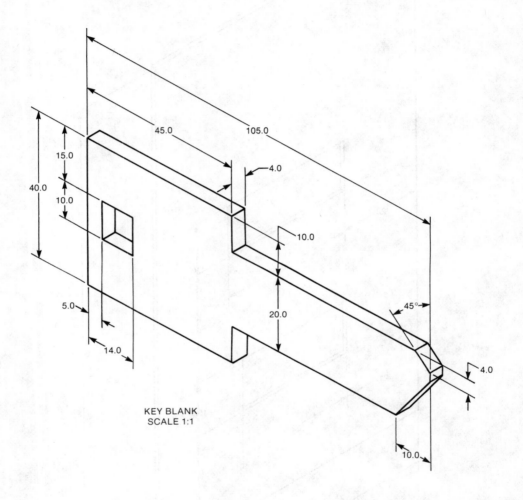

KEY BLANK
SCALE 1:1

Prepare a drawing of the Calculator
Cover. Use an A4 sheet.

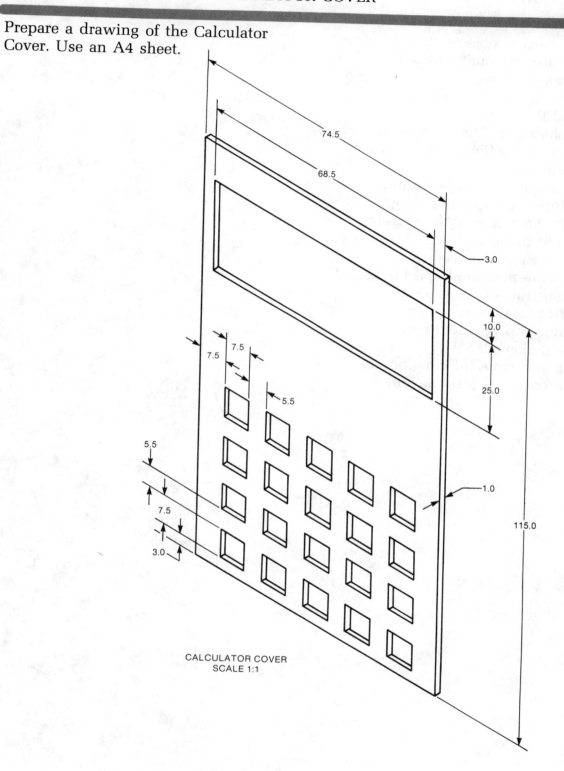

CALCULATOR COVER
SCALE 1:1

- The thickness of objects 12 mm thick or less is usually omitted from drawings.
- Drawings that do not show thickness are called flat layouts.
- The "alphabet of lines" for flat layouts includes construction, object, border, and center lines.
- All the preparation steps should be done before starting a flat layout.
- To locate the center of the working space, light diagonal lines connecting opposite corners are used.
- Distances are measured and marked off before construction lines are drawn.
- Excess line work is removed using an eraser and erasing shield before the lines are darkened.
- The title and scale information are added to complete a flat layout drawing.

CIRCLES AND ARCS

DRAFTING GEOMETRY * PART 1

This unit shows you how to draw simple objects with round shapes. You will discover:

- how to use a compass,
- how to draw flat layouts with circles and arcs,
- how to make the basic geometric construction of circles and arcs.

KEY WORDS

Diameter: The length of a straight line passing through the center of a circle from one side to the other (Figure 14-1). The symbol for diameter is ϕ.

Radius: Half of a diameter. The distance from the center of a circle to the edge (Figure 14-2). The symbol for radius is R.

Arc: A portion of a circle (Figure 14-3).

Fillet: An inside rounded corner (Figure 14-4).

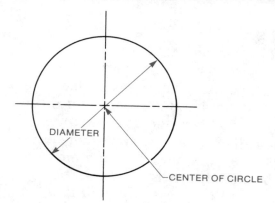

Figure 14-1 *A diameter.*

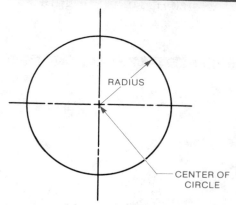

Figure 14-2 *A radius.*

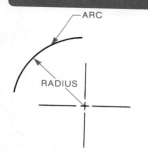

Figure 14-3 *An arc.*

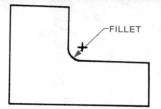

Figure 14-4 *A fillet.*

Round: An outside rounded corner (Figure 14-5).
Bisect: Divide into two equal parts.
Tangent: A line touching but not crossing a curve or surface at one given point.

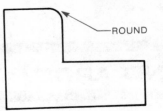

Figure 14-5 A round.

CIRCLES

All circles have a center point where a horizontal center line and a vertical center line cross (Figure 14-1). The distance along a straight line from the center of a circle to the edge (Figure 14-6) is called a *radius*. The *diameter* is the distance from one edge of a circle to the opposite edge. It is measured along a straight line that passes through the center point (Figure 14-7). The diameter tells the size of the circle. The radius is used to tell the size of an *arc* (Figure 14-8).

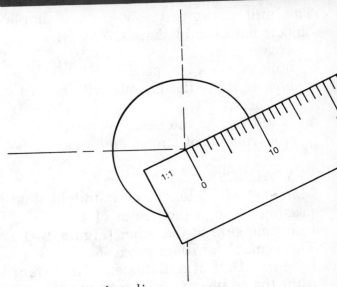

Figure 14-6 A radius.

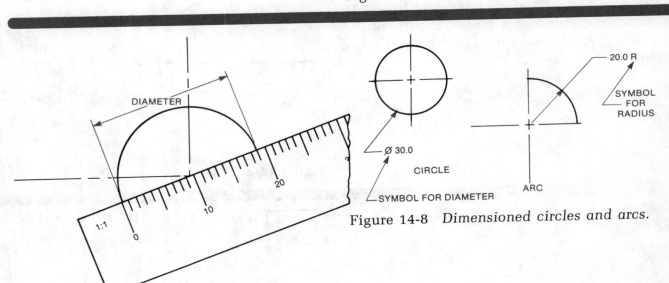

Figure 14-7 A diameter.

Figure 14-8 Dimensioned circles and arcs.

COMPASS PREPARATION

To draw circles, a bow compass is needed. The compass has two legs: one with a needle and one with a lead at the end (Figure 14-9). The needle should have a shouldered point (Figure 14-10).

The slanted edge of the lead should be placed to the outside (Figure 14-11). The lead can be sharpened by rubbing the slanted edge along a sandpaper pad (Figure 14-12). Do not sharpen leads over your drawings (Figure 14-13).

When you use the compass, the needle should be in the board up to the shoulder

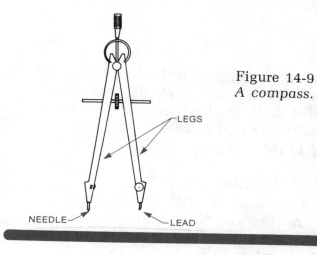

Figure 14-9
A compass.

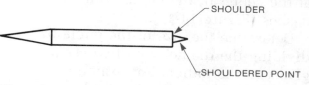

Figure 14-10 *A compass needle.*

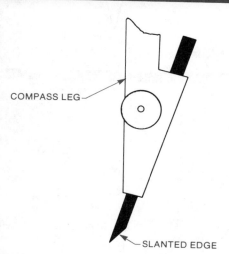

Figure 14-11 *Compass lead.*

Figure 14-12 *Sharpening the compass lead.*

Figure 14-13
Sharpen lead away from drawing.

(Figure 14-14). Adjust the tip of the lead so it is even with the shoulder (Figure 14-15). Both the lead and the needle shoulder should extend 10 mm from the end of the compass leg (Figure 14-16).

DRAWING CIRCLES

To draw a circle, mark the center point by drawing a vertical center line and a horizontal center line. They should cross at the center of the circle with short dashes (Figure 14-17).

Determine the size of the circle by dividing the diameter by two. This will give you the radius. For example:
$$\phi\ 28.0 \div 2 = 14.0\ R.$$

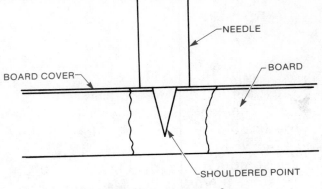

Figure 14-14   Needle in board.

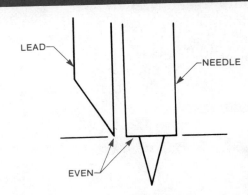

Figure 14-15   Needle and lead even.

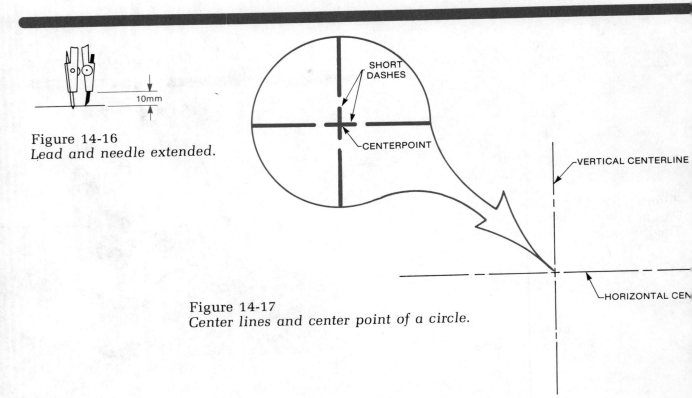

Figure 14-16
Lead and needle extended.

Figure 14-17
Center lines and center point of a circle.

Measure from the center of the circle outward and mark the radius (Figure 14-18). Check the measurement again. Then insert the needle of the compass exactly at the center of the circle (Figure 14-19). The compass should be opened to the size of the radius. Hold the compass at the top and rotate in *one direction only* (Figure 14-20). Rotating the compass in both directions may cause a double line. Continue rotating the compass until the circle line is as thick and as dark as an object line.

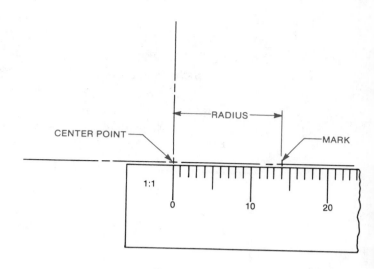

Figure 14-18   *Measure and mark radius.*

Figure 14-20
Compass is held at top.

Figure 14-19
The needle is inserted at the center point.
The compass is opened to radius distance.

Locate the center of an A4 sheet by crossing diagonal lines from the corners. Draw both vertical and horizontal center lines. Draw four circles of different sizes from the center point. The radii should be 20.0 mm, 40.0 mm, 60.0 mm, and 80.0 mm. Darken each circle to the thickness of an object line.

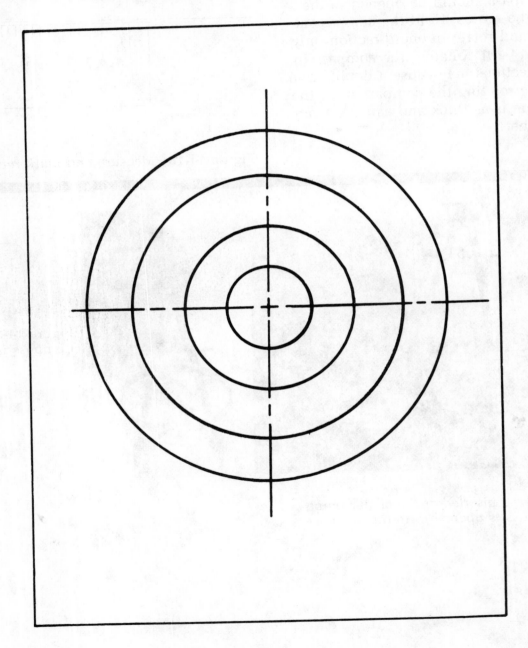

Objects containing circles and arcs require certain steps for drawing. The following are common geometric construction steps for circles and arcs.

BISECTING AN ARC OR LINE

Bisecting means dividing into two equal parts. Figure 14-21 shows the steps for bisecting an arc or line.

Step 1. With the needle of your compass on the end of the line or arc, draw a construction arc. The arc radius should be more than half the distance across the arc or line. Repeat from the other end of the line or arc using the same radius.

Step 2. Draw a line connecting the two points where the construction lines cross.

CONSTRUCTING A CIRCLE TANGENT TO A LINE

A circle is *tangent* to a line when it touches the line at one point only (Figure 14-22).

Step 1. From the point where the circle will touch the line (point of tangency) draw a 90° construction line. The construction line should equal the radius of the circle needed.

Step 2. Place the needle of your compass at point B. With the lead at point A, draw the circle.

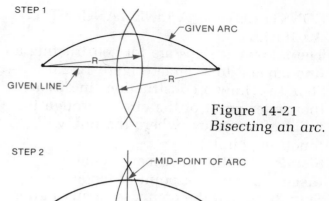

Figure 14-21
Bisecting an arc.

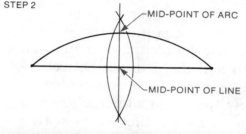

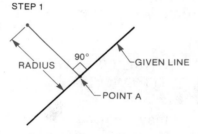

Figure 14-22
Drawing a circle tangent to a line.

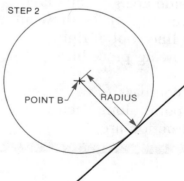

CONSTRUCTING A LINE TANGENT TO A CIRCLE

The following steps are for constructing a line tangent to a circle (Figure 14-23):

Step 1. Draw a construction line from the center point of the circle through the point of tangency (where the line will touch the circle).

Step 2. On the line, mark equal distances from the point of tangency.

Step 3. Open the compass to any radius larger than half the distance between marks. Using each mark as a center point, draw two large construction arcs. Extend the arcs until they cross at two points.

Step 4. Draw the tangent line by connecting the points where the arcs cross.

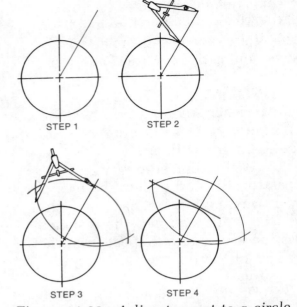

Figure 14-23 A line tangent to a circle.

CONSTRUCTING AN ARC OF A GIVEN RADIUS TANGENT TO TWO LINES NOT AT 90°

The following steps may be used for drawing *fillets* (inside arcs) or *rounds* (outside arcs) (Figure 14-24). To draw an arc tangent to two lines not at right angles use the following procedure (Figure 14-25):

Step 1. Set your compass to the given radius. From any point along each given line, draw a construction arc.

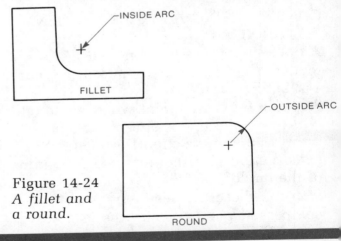

Figure 14-24
A fillet and a round.

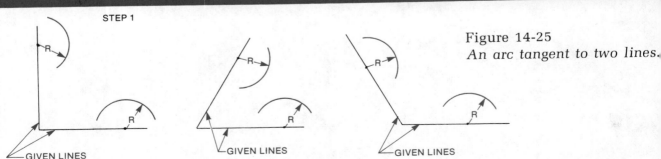

Figure 14-25
An arc tangent to two lines.

Step 2. To locate the center of the needed arc, draw construction lines tangent to the construction arcs and parallel to the two given lines. The point where they intersect is the center of the arc.

Step 3. From the center of the needed arc, draw two lines 90° to the construction lines. This locates the points of tangency on the given lines.

Step 4. Draw the arc between points of tangency.

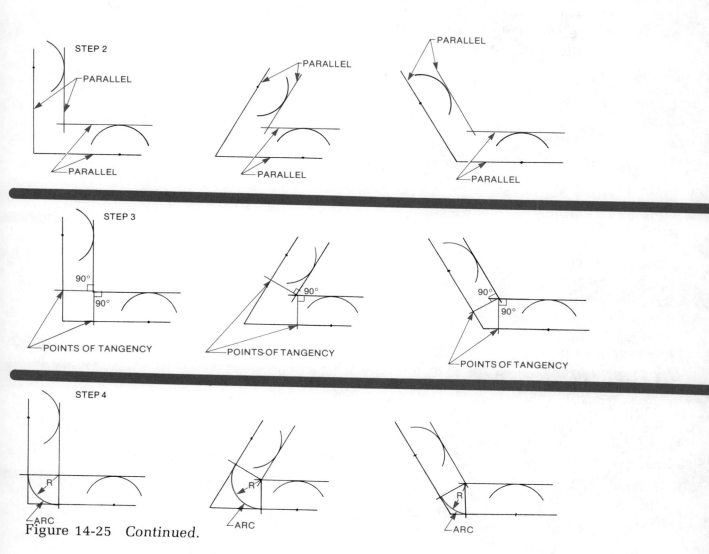

Figure 14-25 Continued.

CONSTRUCTING AN ARC WITH A GIVEN RADIUS TANGENT TO A STRAIGHT LINE AND AN ARC

The following steps apply to fillets or rounds. If you are given a straight line and an arc, construct an arc tangent to both in this way (Figure 14-26):

Step 1. Using the given radius, draw a construction arc from any point along the given arc.

Step 2. Using the center of the given arc, draw a construction arc parallel to the given arc and tangent to the construction arc.

Step 3. Using the given radius, draw another construction arc from any point along the given line.

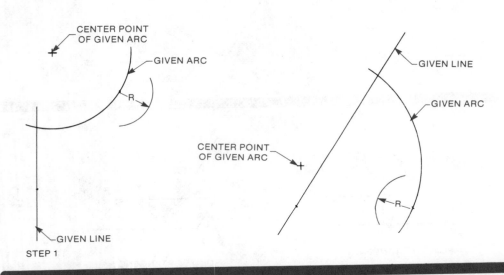

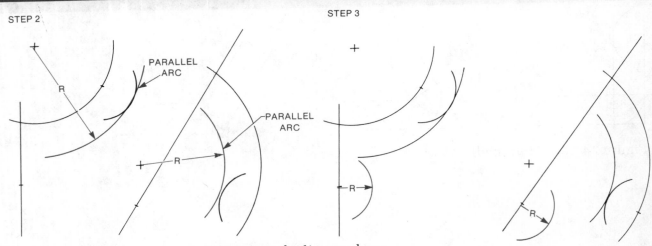

Figure 14-26 *An arc tangent to a straight line and arc.*

Step 4. Draw a construction line parallel to the given line and tangent to the construction arc. Extend it until it crosses the parallel construction arc. This locates the center of the given arc.

Step 5. From this center point, draw a line to the center of the given arc and a line 90° to the given line. These two lines locate the points of tangency on the given arc and line.

Step 6. Draw the arc between points of tangency using the center of the needed arc.

STEP 4

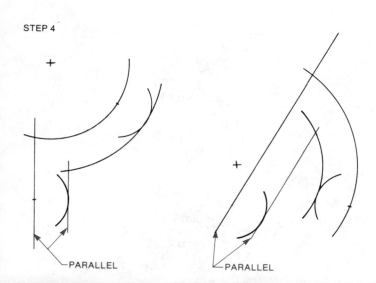

STEP 5

STEP 6

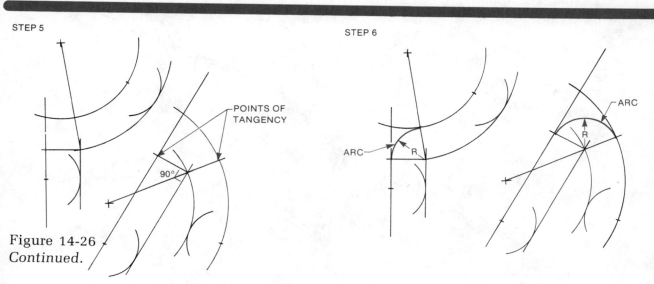

Figure 14-26 Continued.

CONSTRUCTING AN ARC WITH A GIVEN RADIUS TANGENT TO TWO ARCS

The following steps can be used for fillets or rounds. If you are given two arcs, you can draw a third arc tangent to them in this way (Figure 14-27):

Step 1.   Using the given radius, draw one construction arc from any point along each of the given arcs.

Step 2.   Using the given arc centers, draw two construction arcs, one parallel to each given arc and tangent to the construction arcs. Extend the construction arcs until they cross. This locates the center of the arc needed.

Step 3.   To locate tangency points, draw construction lines from the center of the needed arc to the centers of the given arcs.

Step 4.   Draw the arc between points of tangency using the center point for the needed arc.

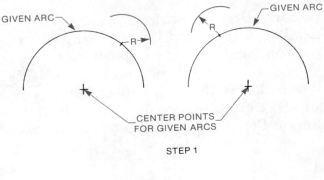

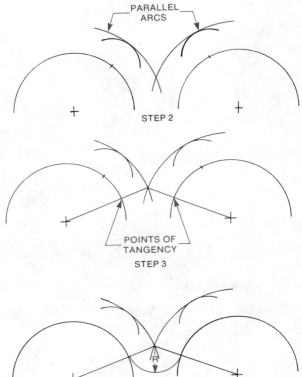

Figure 14-27   *An arc tangent to two arcs.*

On an A4 sheet, draw the geometric
constructions indicated.

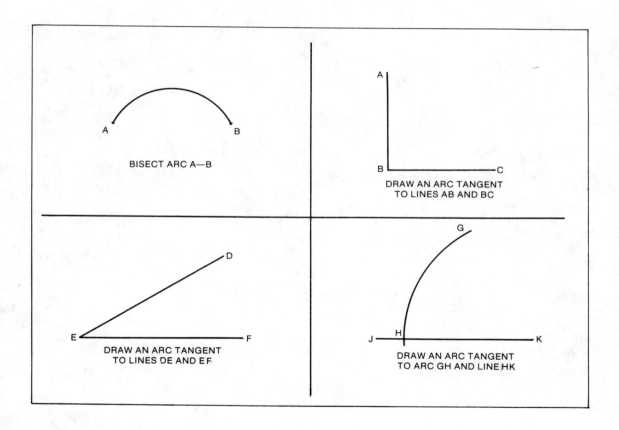

BISECT ARC A—B

DRAW AN ARC TANGENT
TO LINES AB AND BC

DRAW AN ARC TANGENT
TO LINES DE AND EF

DRAW AN ARC TANGENT
TO ARC GH AND LINE HK

This figure shows the half-size drawing with dimensions of the Gasket. Figure 14-28 shows the steps to complete the drawing. Use an A4 sheet.

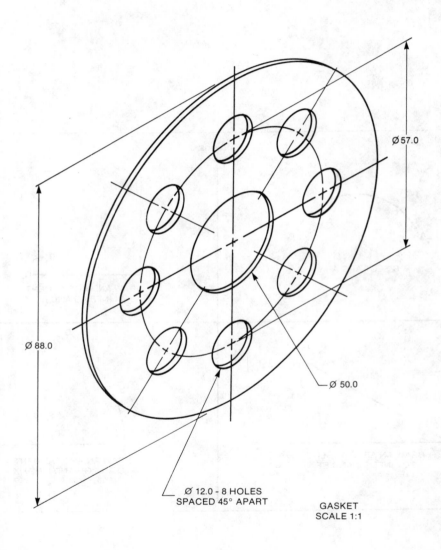

Ø 57.0

Ø 88.0

Ø 50.0

Ø 12.0 - 8 HOLES
SPACED 45° APART

GASKET
SCALE 1:1

Step 1. Draw light diagonal lines connecting corners to locate the center of the working space.

Step 2. Draw both vertical and horizontal center lines passing through the center of the working space. Make sure the short dashes of each center line cross at the center point.

Step 3. Measure and mark the radius distance for the large circles.

Step 4. Draw a light construction circle for each large circle.

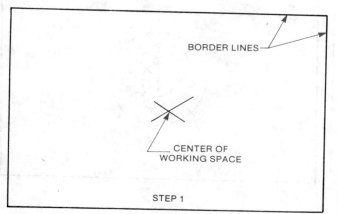

STEP 1

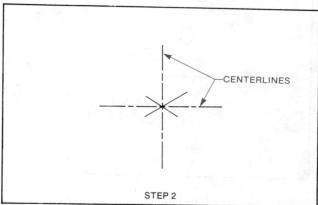

STEP 2

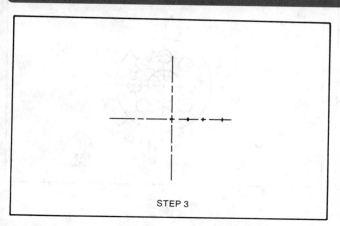

STEP 3

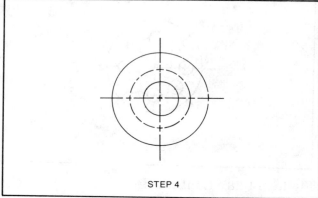

STEP 4

Figure 14-28 *Construction steps for the Gasket.*

Step 5. Draw the center lines for the small circles.

Step 6. Measure and mark the radius distances and draw the small circles. Erase excess line work.

Step 7. Darken all remaining lines.

Step 8. Enter the title and scale information.

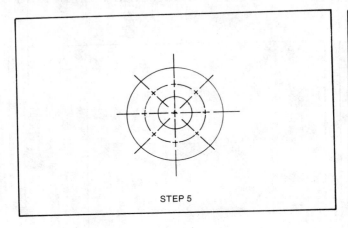

STEP 5

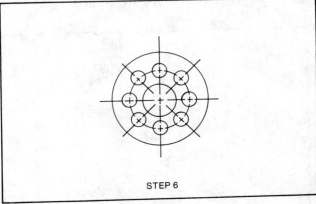

STEP 6

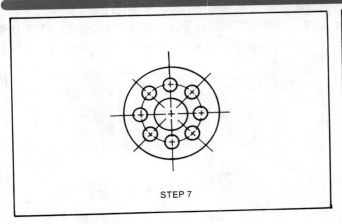

STEP 7

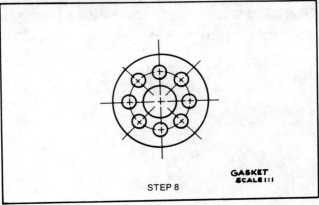

STEP 8

GASKET
SCALE 1:1

Figure 14-28 Continued.

DRAWING PROBLEM 14-2 CHAIN LINK

Using an A4 sheet locate and lay out the
Chain Link in the center of the work
space. Review the steps for drawing
fillets and rounds.

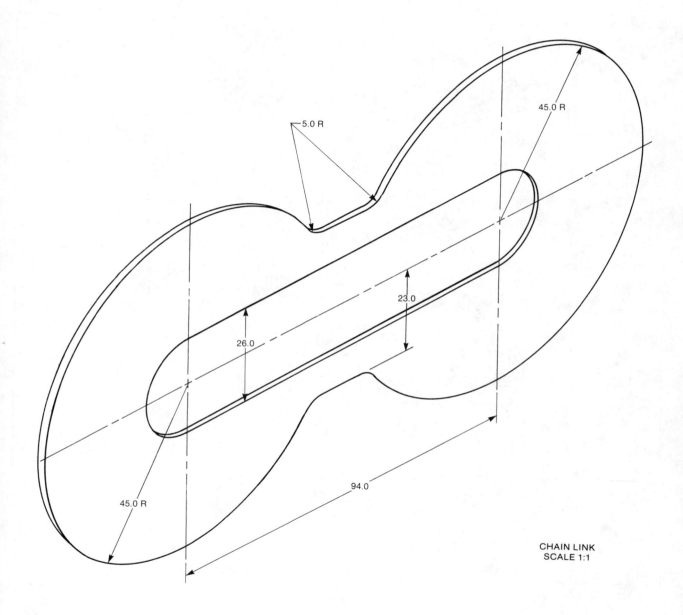

CHAIN LINK
SCALE 1:1

Using an A4 sheet, prepare a flat layout drawing of the Adjuster Plate.

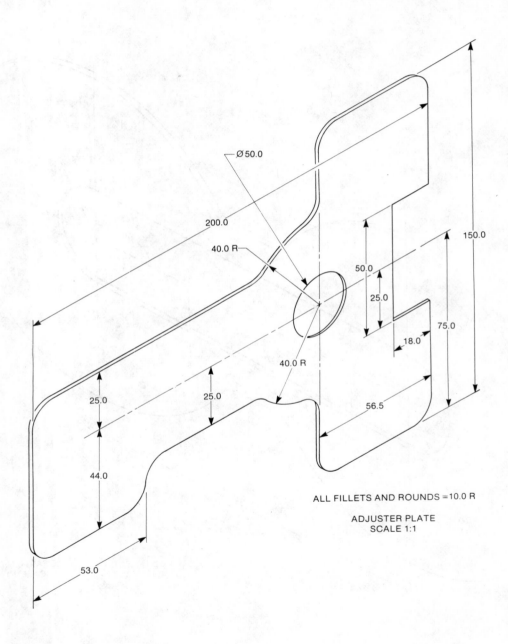

Ø 50.0

200.0

40.0 R

50.0

150.0

25.0

25.0

75.0

18.0

40.0 R

25.0

56.5

44.0

ALL FILLETS AND ROUNDS = 10.0 R

ADJUSTER PLATE
SCALE 1:1

53.0

Using an A4 sheet, prepare a flat layout
of the Template.

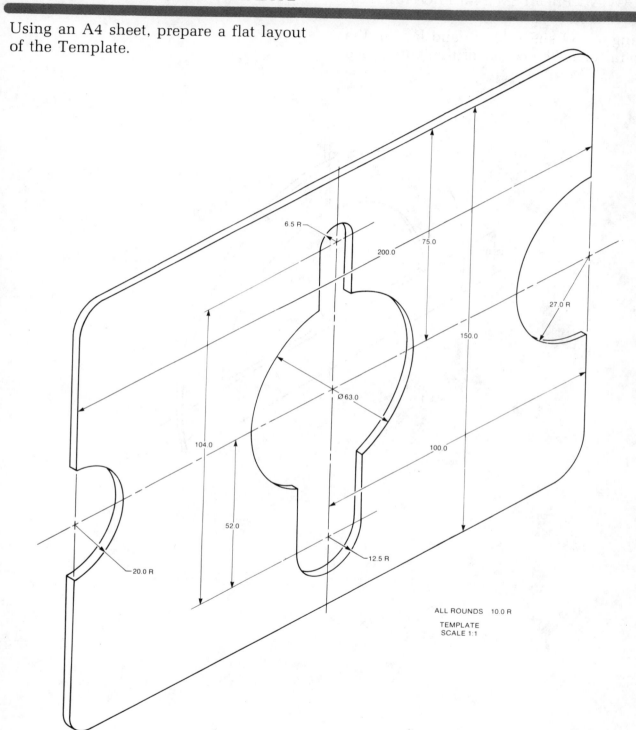

6.5 R

200.0

75.0

150.0

27.0 R

Ø 63.0

104.0

100.0

52.0

12.5 R

20.0 R

ALL ROUNDS   10.0 R

TEMPLATE
SCALE 1:1

Using an A4 sheet, locate and lay out the
Protractor in the center of the work space.

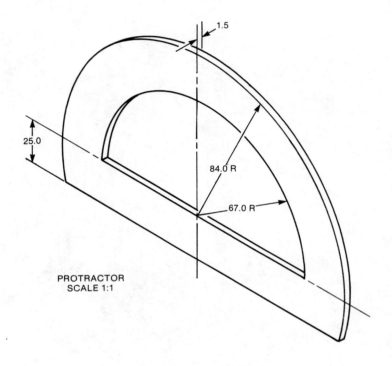

1.5

25.0

84.0 R

67.0 R

PROTRACTOR
SCALE 1:1

Using an A4 sheet, prepare a flat layout
drawing of the Gate Brace.

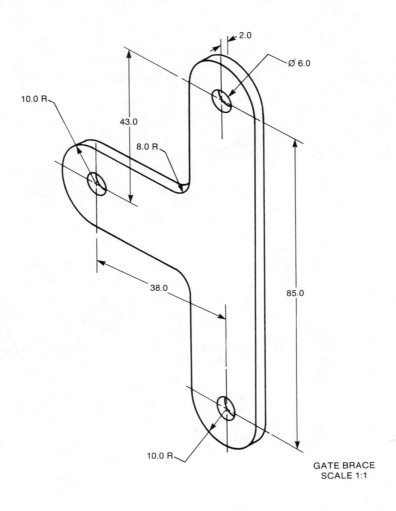

GATE BRACE
SCALE 1:1

Use an A4 sheet. Prepare a flat layout
drawing of the Hook Latch.

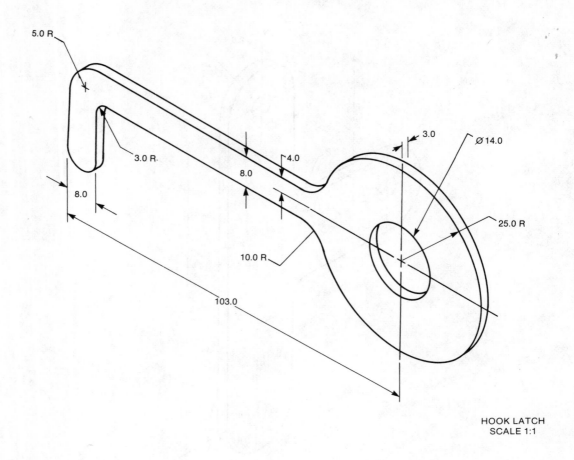

5.0 R

3.0 R

8.0

8.0

3.0

4.0

Ø 14.0

25.0 R

10.0 R

103.0

HOOK LATCH
SCALE 1:1

Prepare a flat drawing of the Receptacle
Plate. Use an A4 sheet.

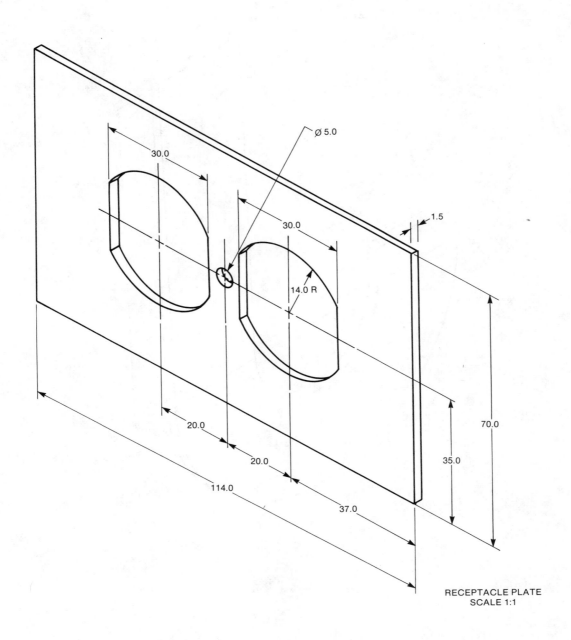

Ø 5.0

30.0

30.0

1.5

14.0 R

20.0

20.0

70.0

35.0

114.0

37.0

RECEPTACLE PLATE
SCALE 1:1

Using an A4 sheet, prepare a flat layout
drawing of the Block Letter.

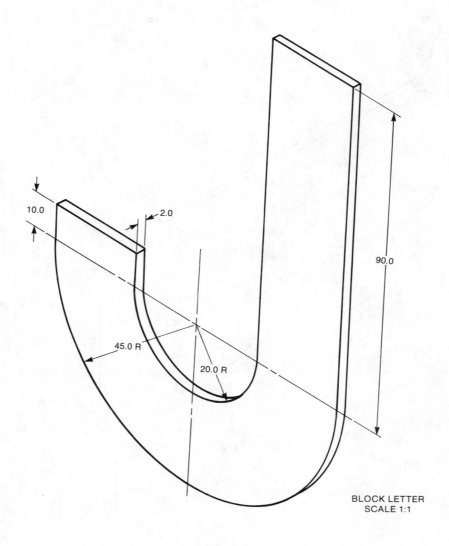

10.0

2.0

90.0

45.0 R

20.0 R

BLOCK LETTER
SCALE 1:1

Use an A4 sheet to prepare a flat layout
drawing of the Wrench.

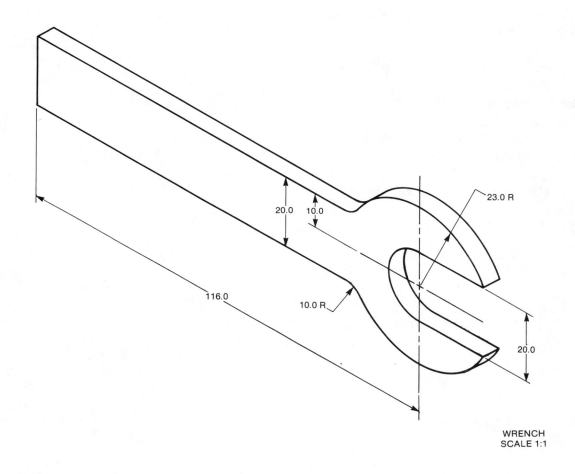

23.0 R

20.0   10.0

116.0

10.0 R

20.0

WRENCH
SCALE 1:1

- A diameter is the distance along a straight line passing through the center of a circle from one end to the other.
- A radius is half of the diameter.
- Circles are drawn with a vertical and a horizontal center line passing through the center point.
- A compass lead is sharpened with the slant on the outside.
- The tip of the compass lead is aligned with the shoulder of the needle.
- Objects with circles and arcs require geometric construction steps to draw them.
- Arcs that curve to the outside are called rounds.
- Arcs that curve to the inside are called fillets.

GEOMETRIC SHAPES

DRAFTING GEOMETRY * PART 2

This unit shows you how to draw common shapes. You will learn:
● the common geometric shapes,
● how to draw the common geometric shapes.

KEY WORDS
Square: A rectangle with four equal sides.
Rectangle: A four-sided figure with parallel sides (Figure 15-1).
Triangle: A three-sided figure.
Equilateral Triangle: A triangle with three equal sides and three equal angles (Figure 15-2).
Hexagon: A six-sided figure (Figure 15-3).
Octagon: A figure that has eight sides (Figure 15-4).

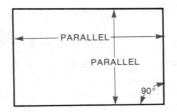

Figure 15-1   *A rectangle.*

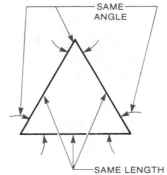

Figure 15-2   *An equilateral triangle.*

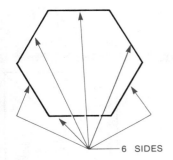

Figure 15-3   *A hexagon.*

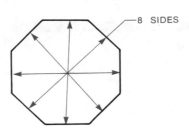

Figure 15-4   *An octagon.*

GEOMETRIC SHAPES

A drafter must know and draw geometric shapes. Rectangles, squares, triangles, hexagons, and octagons are very common geometric shapes (Figure 15-5). These shapes are found in the construction of many things.

TOOLS FOR GEOMETRIC CONSTRUCTION

To draw geometric shapes a drafter uses a compass, scale, T-square, triangles, a 3H pencil for construction, and a 2H pencil for darkening.

DRAWING A SQUARE

A *square* (Figure 15-6) can be drawn several ways. The method used depends upon what you know about the square. Three methods are commonly used.

When you know the length of the side (Figure 15-7) follow this method:

Step 1. Measure, mark, and draw the length of the side.

Step 2. Draw a light diagonal construction line from one end of the length at a 45° angle.

Step 3. From the other end of the length, draw a vertical (90°) line and extend it until it intersects the 45° line.

Step 4. Close off the square with lines parallel to the two sides already drawn.

Figure 15-5
Common shapes. (Joseph T. Ryerson & Son, Inc.)

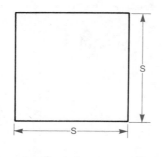

Figure 15-6
The square.

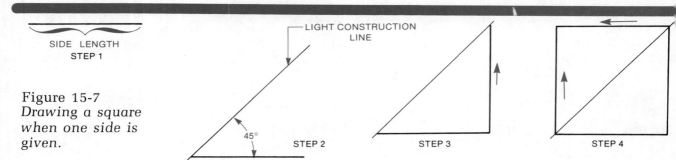

SIDE LENGTH
STEP 1

Figure 15-7
Drawing a square when one side is given.

LIGHT CONSTRUCTION LINE

45°
STEP 2

STEP 3

STEP 4

Another method to use if you know the length of the side is this (Figure 15-8):

Step 1.   Draw a circle with a radius equal to one-half the side length.

Step 2.   Draw four light lines, two horizontal, touching the top and bottom of the circle, and two vertical, touching the sides. Then darken the lines.

If, instead of the length, you know the distance across corners (Figure 15-9), follow this method (Figure 15-10):

Step 1.   Draw a circle with a radius half the distance across corners.

Step 2.   Through the center of the circle, draw two lines at 45° to horizontal. Extend the lines until they touch the outside of the circle.

Step 3.   Connect the four points where the diagonals touch the circle with two horizontal and two vertical lines. Darken the lines.

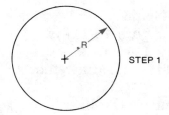

STEP 1

R = HALF SIDE LENGTH

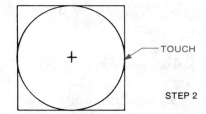

TOUCH

STEP 2

Figure 15-8
Alternate method of drawing a square when one side is given.

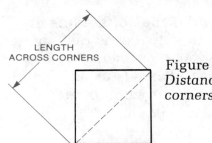

LENGTH ACROSS CORNERS

Figure 15-9
Distance across corners of a square.

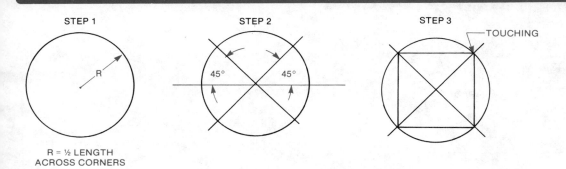

STEP 1

R = ½ LENGTH ACROSS CORNERS

STEP 2

45°   45°

STEP 3

TOUCHING

Figure 15-10   *Drawing a square when distance across corners is given.*

DRAWING A RECTANGLE

Drawing a rectangle (Figure 15-11) is similar to constructing a square.

Step 1. Measure, mark, and draw the horizontal length.

Step 2. From one end of the horizontal line, measure, mark, and draw the vertical length.

Step 3. Close off the rectangle with another vertical and horizontal.

DRAWING AN EQUILATERAL TRIANGLE

An *equilateral triangle* has three equal sides and is constructed in the following way (Figure 15-12):

Step 1. Measure one side, mark and draw it.

Step 2. Open a compass to the length of the side drawn.

Step 3. Using the ends of the line as center points, draw two construction arcs of the same radius that intersect.

Step 4. Connect the ends of the line to the point of the intersection of arcs. Then darken the lines.

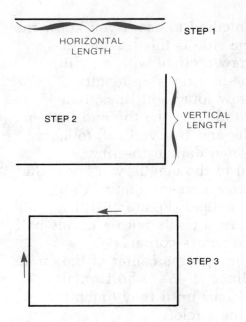

Figure 15-11   *Drawing a rectangle.*

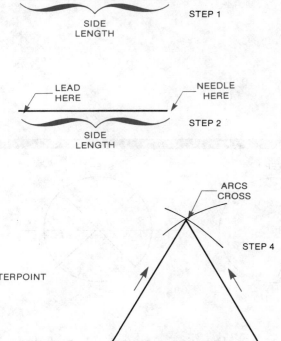

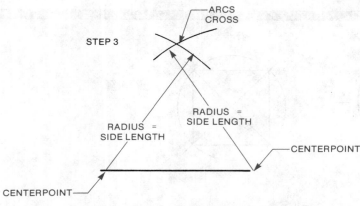

Figure 15-12   *Drawing an equilateral triangle.*

DRAWING A HEXAGON

There are two methods of drawing a *hexagon*. The method you use depends upon what you know about the hexagon (Figure 15-13).

When you know the distance across flats, use this method (Figure 15-14):

Step 1. Draw a circle with a diameter equal to the distance across flats.

Step 2. Draw two horizontal lines tangent to the top and bottom of the circle.

Step 3. Complete the hexagon by drawing four lines also tangent to the circle and 30° from the vertical. Then darken the lines of the completed hexagon.

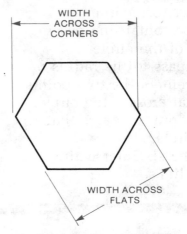

Figure 15-13   *A hexagon.*

Figure 15-14   *Drawing a hexagon when distance across flats is known.*

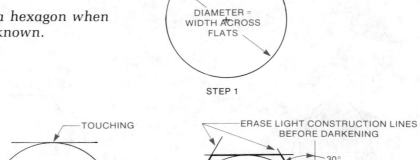

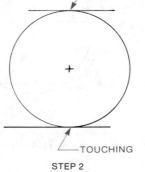

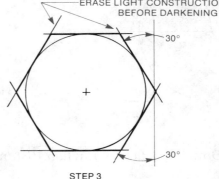

If you know the distance across corners, draw the hexagon in this way (Figure 15-15):

Step 1.   Draw a circle with a diameter equal to the width across corners.

Step 2.   Draw a horizontal line passing through the center of the circle.

Step 3.   Set a compass to the radius of the circle. Use as centers the two points where the horizontal crosses the circle. From each center, draw two arcs that pass through the circle.

Step 4.   Connect the six intersecting points. Darken the lines.

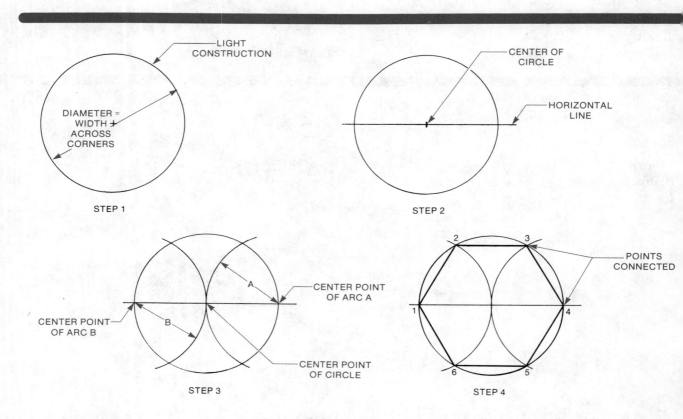

Figure 15-15   *Drawing a hexagon when distance across corners is known.*

DRAWING AN OCTAGON

An octagon is a figure with eight equal sides. It is drawn in this way (Figures 15-16 and 15-17).

Step 1. Draw a circle with a diameter equal to the width across flats.

Step 2. Draw two vertical and two horizontal lines each tangent to the circle.

Step 3. Draw four lines 45° to the horizontals and verticals. Darken the lines to complete the octagon.

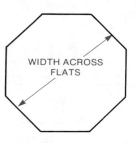

Figure 15-16 *An octagon.*

Figure 15-17 *Drawing an octagon.*

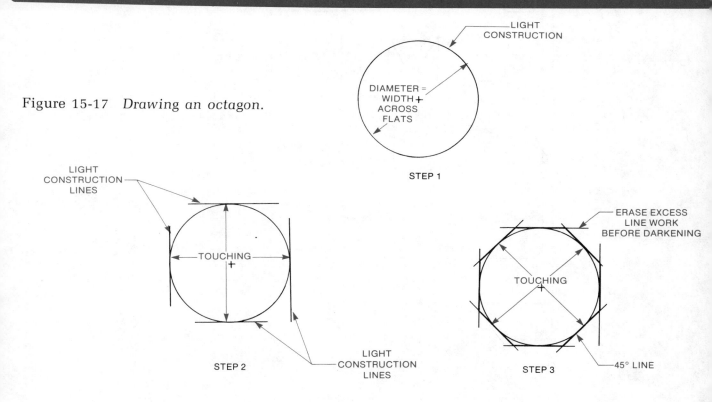

On an A4 sheet draw the Speed Limit
Sign. Center your drawing. Letter the title
and scale information in the lower right-
hand corner.

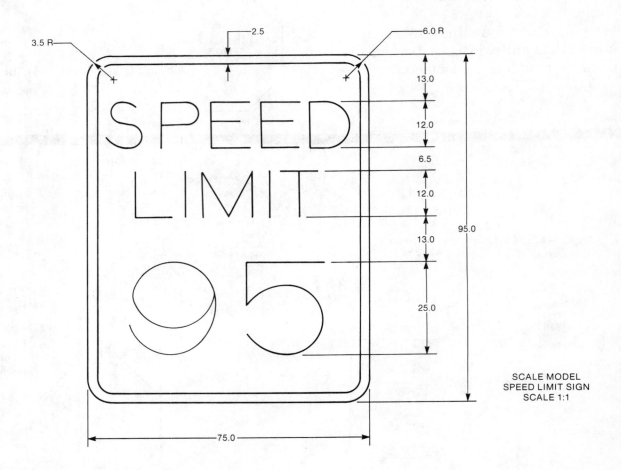

SCALE MODEL
SPEED LIMIT SIGN
SCALE 1:1

On an A4 sheet draw the Yield Sign in
the center of the working space. Review
the procedure for drawing an equilateral
triangle before starting. Add the title and
scale.

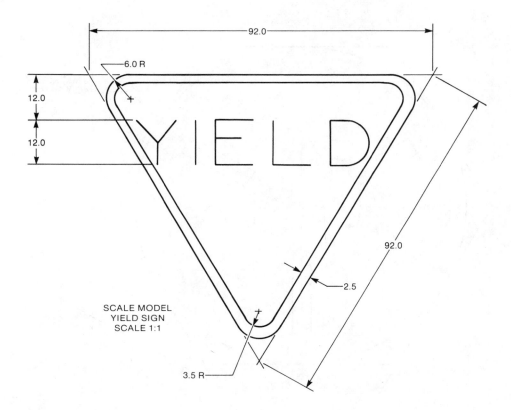

SCALE MODEL
YIELD SIGN
SCALE 1:1

Use an A4 sheet. Locate the Head Detail
in the center of the working space.

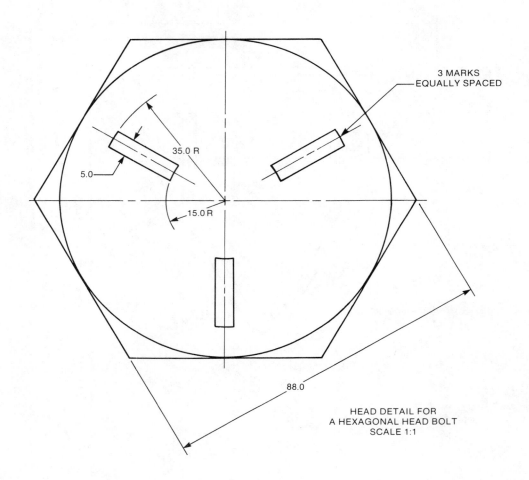

3 MARKS
EQUALLY SPACED

35.0 R

5.0

15.0 R

88.0

HEAD DETAIL FOR
A HEXAGONAL HEAD BOLT
SCALE 1:1

On an A4 sheet draw the Stop Sign. Add
the title and scale. Review procedures for
drawing a hexagon before starting.

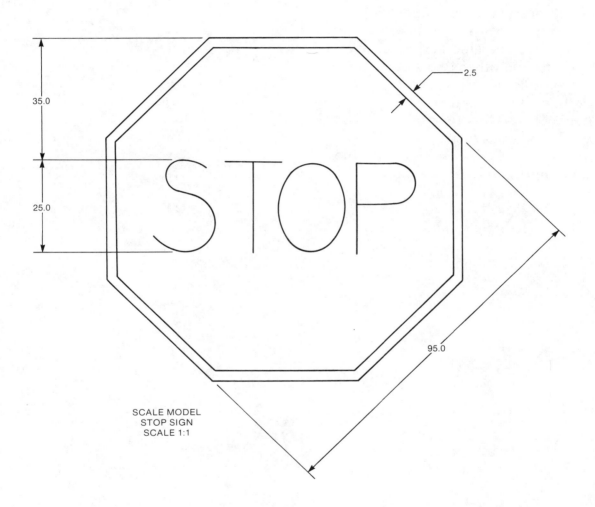

35.0

2.5

25.0

95.0

SCALE MODEL
STOP SIGN
SCALE 1:1

- Squares, rectangles, triangles, hexagons, and octagons are shapes found in many things.
- Squares can be drawn using the distance along the side or the distance across the corners.
- A rectangle is drawn by making two measurements: length and width.
- An equilateral triangle is drawn by measuring one side and using two arcs to locate the other two sides.
- A hexagon can be drawn by two methods. One method uses the width across flats. The other method uses the width across corners.
- An octagon is drawn by using lines touching the edges of a circle.

CIRCULAR MEASUREMENT

This unit shows you how to measure and draw angles. You will discover:
- how to use a protractor for measuring angles,
- how to measure and draw angles accurately.

KEY WORDS
Protractor: An instrument used to measure angles (Figure 16-1).
Vertex: The point where the two sides of an angle meet (Figure 16-2).
Legs: Lines that form the sides of the angle (Figure 16-2).

UNITS FOR MEASURING ANGLES
The units of measurement for angles are called *degrees.* Figure 16-3 shows the symbol used to indicate degrees. A complete circle contains 360 degrees (Figure 16-4).

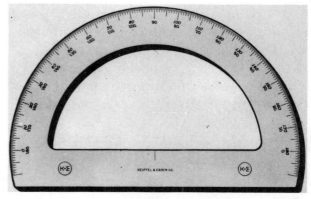

Figure 16-1   *A protractor. (Keuffel & Esser Co.)*

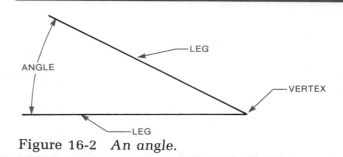

Figure 16-2   *An angle.*

Figure 16-3   *The degree symbol.*

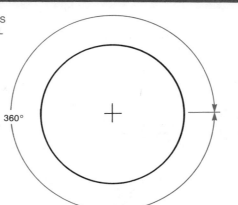

Figure 16-4
A circle contains 360 degrees.

NAMES OF ANGLES

Angles are named by their size measured in degrees:

Acute Angles. An acute angle is any angle that measures less than 90 degrees (Figure 16-5).

Obtuse Angles. An obtuse angle measures more than 90 degrees but less than 180 degrees (Figure 16-5).

Reflex Angles. A reflex angle is any angle that measures more than 180 degrees (Figure 16-5).

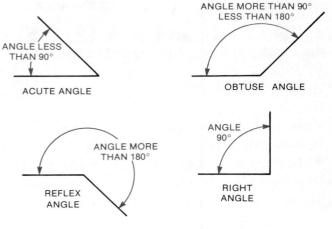

Figure 16-5   *The names of angles.*

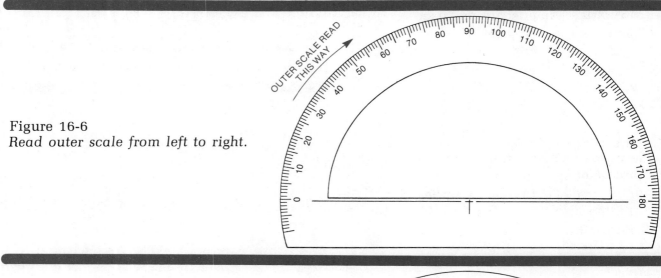

Figure 16-6
Read outer scale from left to right.

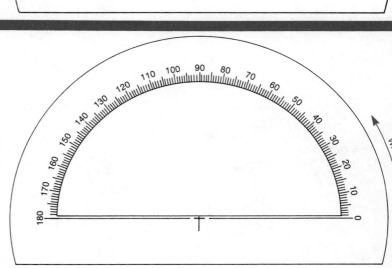

Figure 16-7
Read inner scale from right to left.

Right Angles. An angle that measures 90 degrees is a right angle (Figure 16-5).

THE PROTRACTOR
Angles are measured with a *protractor* (Figure 16-1). A protractor has an inner scale and an outer scale. The outer scale is read from left to right (Figure 16-6). The inner scale is read from right to left (Figure 16-7). Each mark represents one degree (Figure 16-8). Every protractor has a mark at the center of the base line. This is called the *vertex.*

MEASURING ANGLES
To measure an angle, place the vertex of the protractor directly over the vertex of the angle. Place one *leg* of the angle on the base line. Figure 16-9 shows the vertex and base line. Figure 16-10 shows

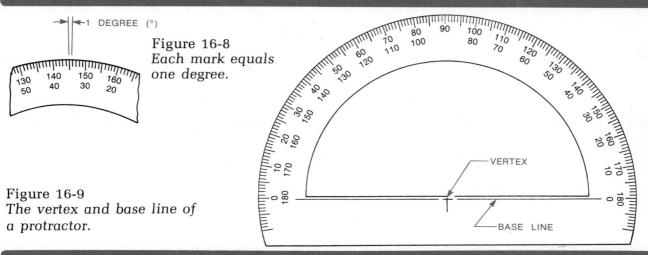

Figure 16-8
Each mark equals one degree.

Figure 16-9
The vertex and base line of a protractor.

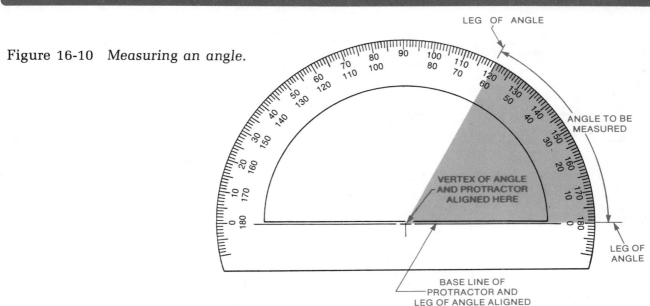

Figure 16-10   *Measuring an angle.*

an angle to be measured (shaded area).
Align the bottom leg along the base line
of the protractor. Place the vertex of the
protractor on the vertex of the angle. You
can then read the degrees where the
other leg of the angle crosses the
protractor.

DRAWING ANGLES
Draw one leg of the angle. Then place
the vertex of the protractor at the point
where the vertex of the angle is to be.
Measure and mark off the angle (Figure
16-11). To complete the angle, draw a
straight line from the point marked off to
the vertex. Figure 16-12 shows the
completed angle.

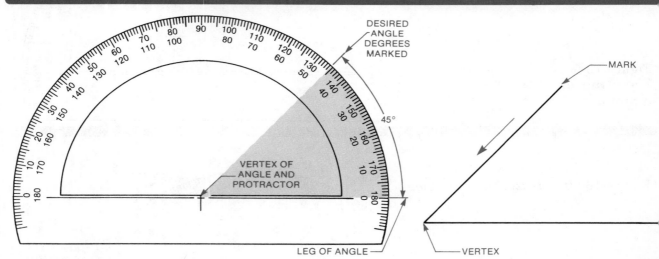

Figure 16-11 *Laying out an angle.*

Figure 16-12 *Angle completed.*

Draw each of the five angles shown in
the figure. Be sure to align your protractor correctly.

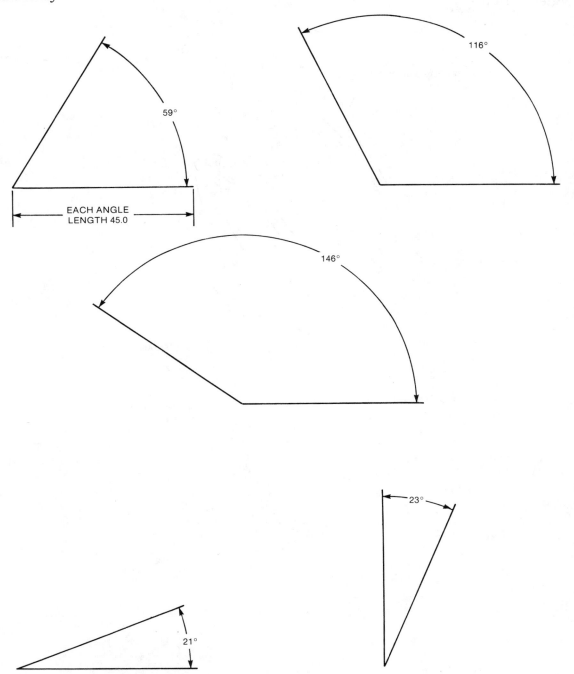

Use an A4 sheet size. Prepare a flat
layout of the Positioning Disk. Lay out
the angles carefully.

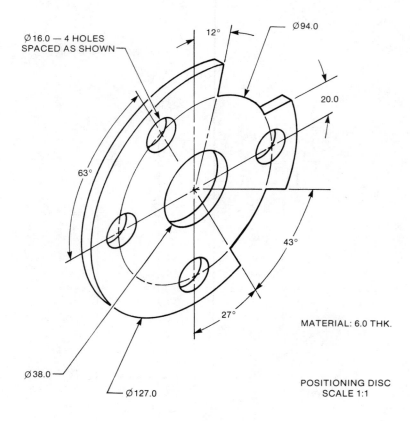

Ø16.0 — 4 HOLES
SPACED AS SHOWN

12°

Ø94.0

20.0

63°

43°

27°

MATERIAL: 6.0 THK.

Ø38.0

Ø127.0

POSITIONING DISC
SCALE 1:1

Use an A3 sheet. Prepare a flat layout of the Clutch Arm.

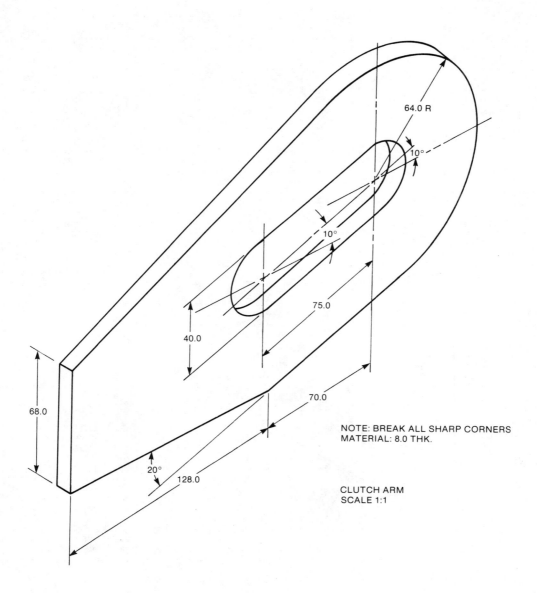

64.0 R

10°

10°

75.0

40.0

70.0

68.0

20°

128.0

NOTE: BREAK ALL SHARP CORNERS
MATERIAL: 8.0 THK.

CLUTCH ARM
SCALE 1:1

Use an A4 sheet. Prepare a flat layout of
the Drill Index.

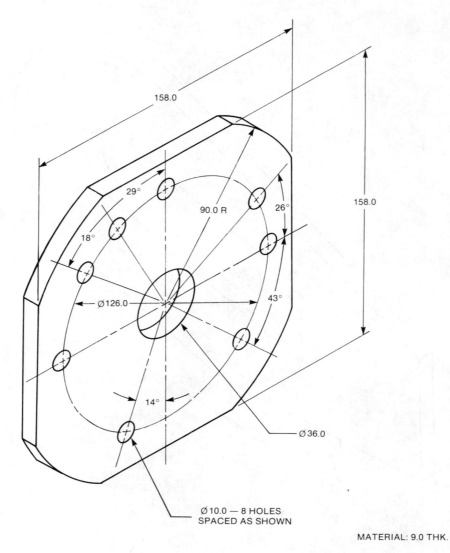

158.0

158.0

29°

90.0 R

26°

18°

43°

Ø126.0

14°

Ø 36.0

Ø 10.0 — 8 HOLES
SPACED AS SHOWN

MATERIAL: 9.0 THK.

DRILL INDEX
SCALE 1:1

- Circular measurement includes the use of angles.
- An angle is the space between two straight lines that meet at a point.
- The units of measurement for angles are degrees.
- Right angles are angles that measure 90 degrees.
- Acute angles measure less than 90 degrees.
- Obtuse angles measure more than 90 degrees but less than 180 degrees.
- Reflex angles measure more than 180 degrees.
- Protractors are used to measure angles.
- To measure or to draw angles, you must align the vertex of the protractor and the vertex of the angle.
- When measuring angles, the base line of the protractor is aligned with one leg of the angle.

SIMPLE OBJECTS

MULTIVIEW DRAWINGS * PART 1

This unit shows you how to draw objects that are not flat. You will discover:
- how to draw objects that cannot be shown by flat layouts,
- how to draw three-dimensional objects correctly.

KEY WORDS

Multiview Drawings: Drawings that show more than one view of an object. Figure 17-1 shows three views of the object pictured in Figure 17-2.

Orthographic View: A drawing that shows a side of an object viewed directly from 90° (Figure 17-2).

Projection: Transferring distances on a drawing using light lines (Figure 17-2).

MULTIVIEW DRAWINGS

Not all objects are simple enough to be drawn in one-view (Figure 17-3). The complex features of some objects require two, three, or more views of the object to best show its shape. The number of views required usually varies with the complexity of the object.

VISUALIZATION

The three views most often used in drawing are the front, top, and right side views (Figure 17-4).

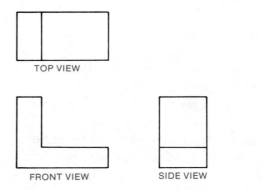

Figure 17-1   *A multiview drawing.*

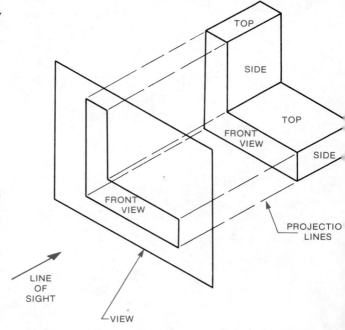

Figure 17-2   *An orthographic view.*

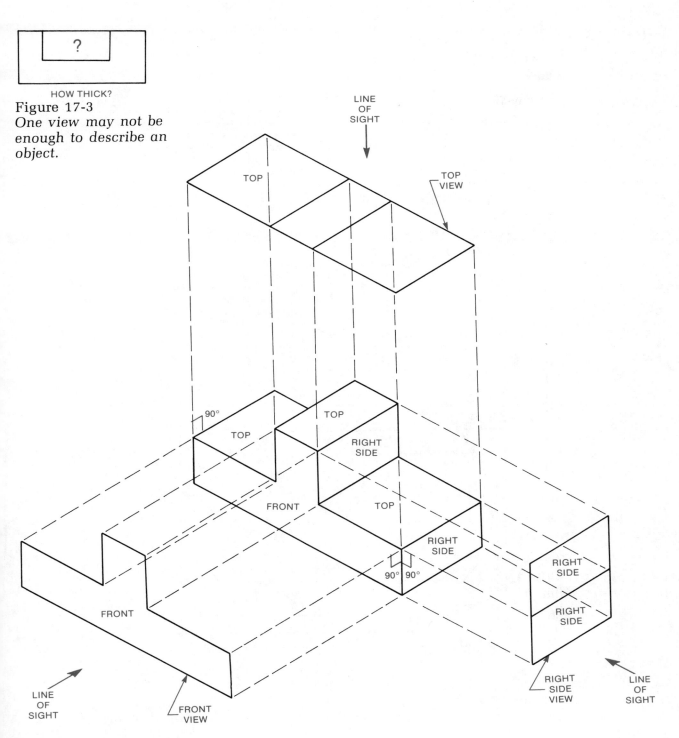

HOW THICK?

Figure 17-3
One view may not be
enough to describe an
object.

Figure 17-4   The three basic views.

SHEET LAYOUT OF BASIC ORTHOGRAPHIC VIEWS

In a multiview drawing, the basic drawings are arranged in a certain way (Figure 17-5):

Front View: Drawn in lower left-hand corner of sheet.

Top View: Drawn directly above the front view.

Right Side View: Drawn directly to the right of the front view.

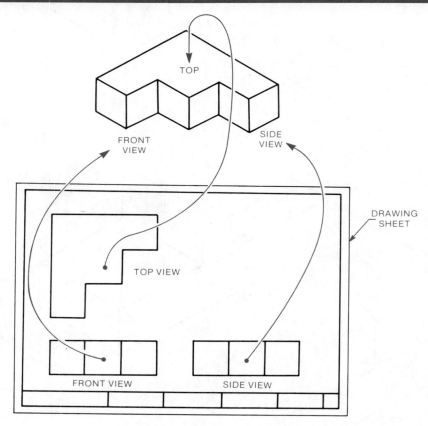

TRADITIONAL LAYOUT OF THE THREE
ORTHOGRAPHIC VIEWS

Figure 17-5
The front, top, and right side views on a drawing sheet.

SPACING THE VIEWS

Spaces between views, and spaces between views and border lines, should be approximately equal (Figure 17-6).

PROJECTION LINES

To keep each view in line with the others, and to avoid measuring the same dimension over and over, the drafter uses light construction lines called projection lines (Figure 17-7). The use of projection lines to construct the views is called *projection*.

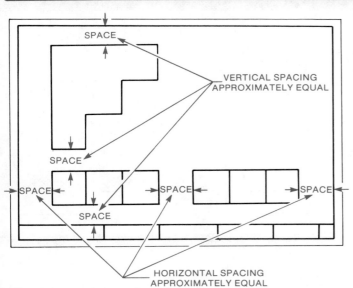

Figure 17-6   *Spacing the views.*

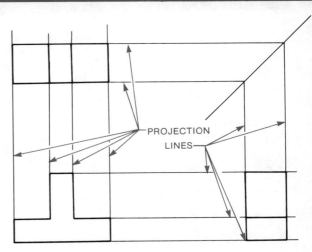

Figure 17-7   *Projection lines.*

CONSTRUCTING VIEWS BY PROJECTION

To construct a multiview drawing by projection follow this procedure (Figure 17-8):

Step 1. Draw and darken the front view.

Step 2. Project the length upward from the front view.

Step 3. Project the height to the right from the front view.

Step 4. Measure and draw the width for the side view.

Step 5. Darken the side view.

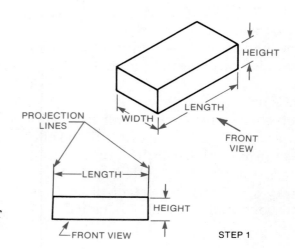

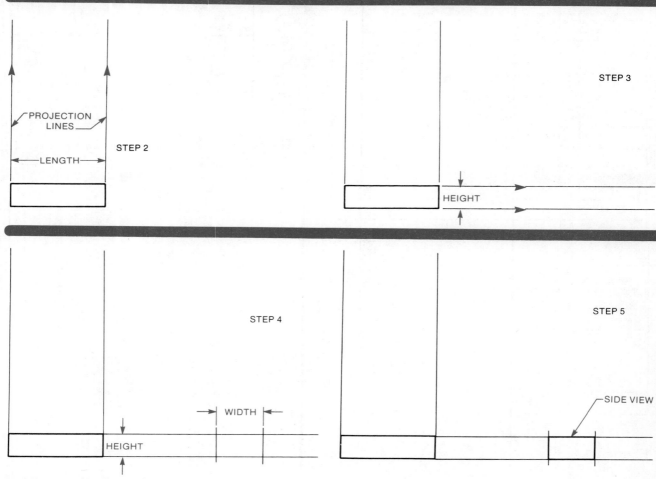

Figure 17-8 *Projecting the views.*

Step 6. Project the width up from the side view.

Step 7. To turn the width towards the top view, draw a 45 degree construction line (called a miter line).

Step 8. Project the width from the miter line to close off the top view.

Step 9. Darken the top view. Note: the projection lines may be left on the drawing if they are extremely light.

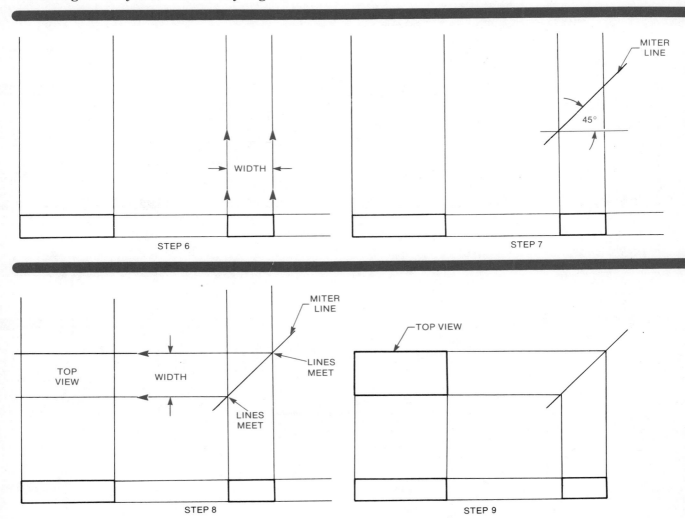

Figure 17-8 *Continued.*

PLACING THE MITER LINE

Place the miter line carefully. If the miter line is too low, the top view will be too close to the front view (Figure 17-9). If the miter line is placed too high, the top view will be too close to the top border (Figure 17-10).

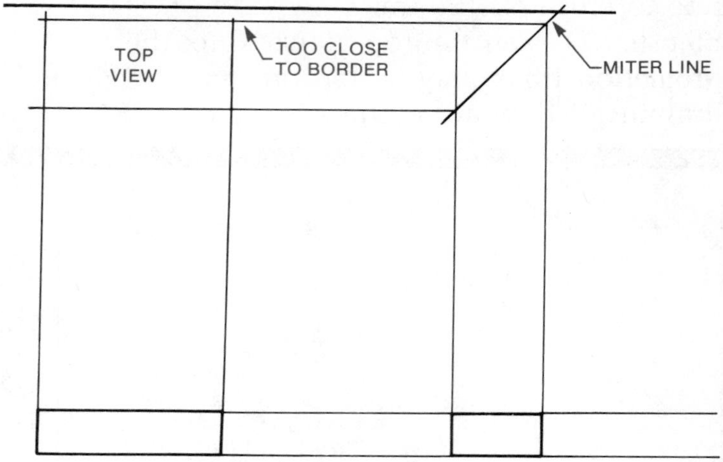

Figure 17-9   *Miter line too low.*

Figure 17-10   *Miter line too high.*

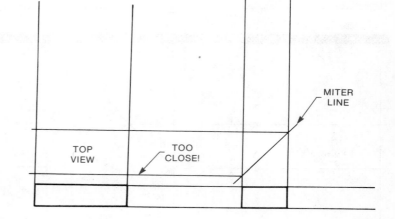

ALTERNATE VIEWS

Sometimes the front, top and right side views may not be enough to show the entire object. Alternate views such as the bottom, back, and left side views may be needed (Figure 17-11). Note: dashed lines in Figure 17-11 represent hidden edges not visible in the view. Alternate views are located as follows:

Bottom View: Located directly below the front view.

Left Side View: Located directly to the left of the front view.

Rear View: Shown directly to the left of the left side view.

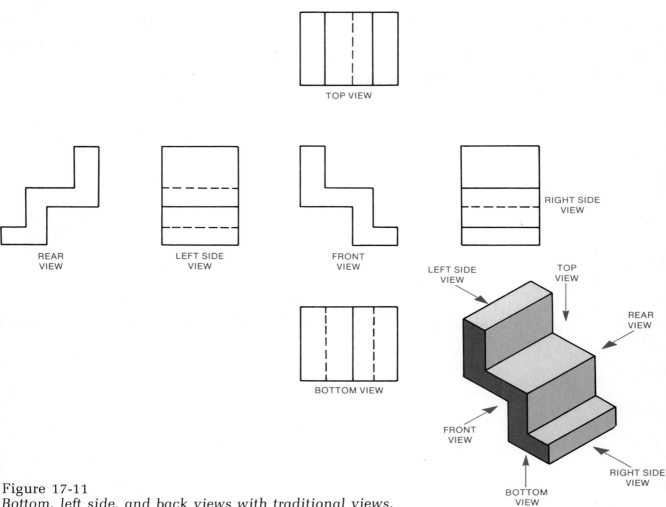

Figure 17-11
Bottom, left side, and back views with traditional views.

SELECTING THE PROPER VIEWS

To show the object clearly is the purpose of any drawing. Some objects require one view, as flat layouts do (Figure 17-12). Some objects may require two views (Figure 17-13). Most objects require three views (Figure 17-14). Other objects may require more than three views.

When selecting views, the drafter selects the view that shows the object's shape most clearly. This view is drawn as the front view. Additional views such as top, right, left, bottom, or rear views are then selected only if they help describe the object. As a general rule, the drafter should draw enough views to show each feature clearly once.

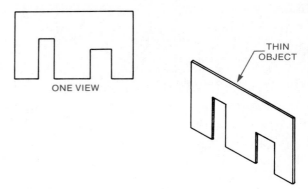

Figure 17-12 *A one-view drawing.*

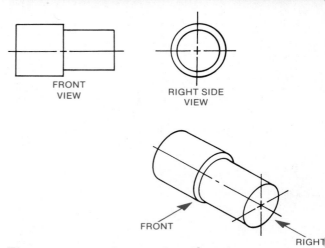

Figure 17-13 *A two-view drawing.*

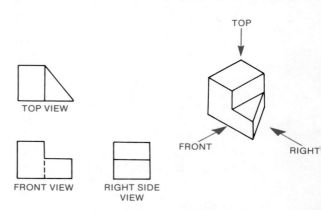

Figure 17-14 *A three-view drawing.*

On an A3 sheet, complete the missing view of each of the five problems. Pay close attention to the arrow marked *front view* in the illustration of the object.

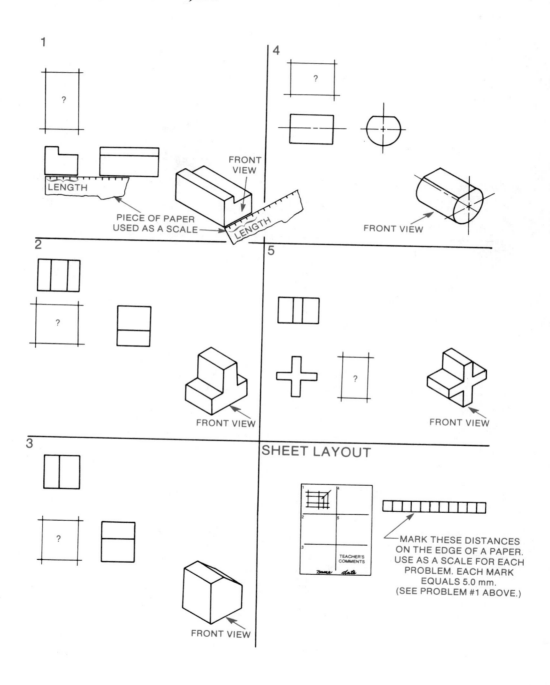

SHEET LAYOUT

MARK THESE DISTANCES ON THE EDGE OF A PAPER. USE AS A SCALE FOR EACH PROBLEM. EACH MARK EQUALS 5.0 mm. (SEE PROBLEM #1 ABOVE.)

On an A3 sheet, complete the missing view of each of the five problems. Supply either the front, top, or right side view as indicated.

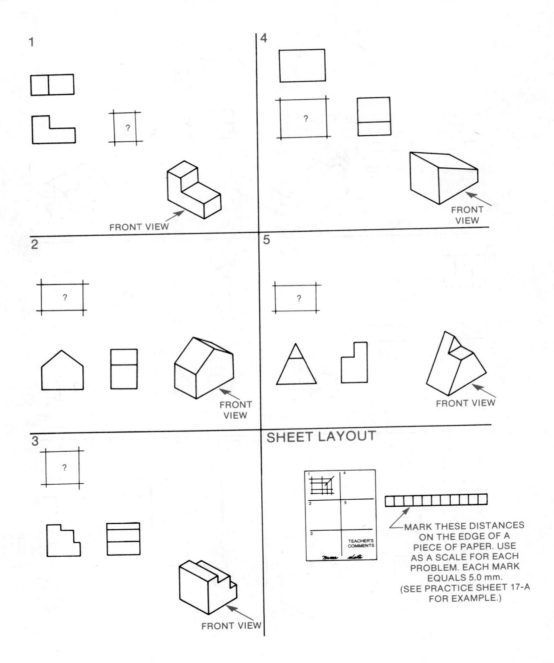

1

FRONT VIEW

4

FRONT VIEW

2

FRONT VIEW

5

FRONT VIEW

3

FRONT VIEW

SHEET LAYOUT

TEACHER'S COMMENTS

MARK THESE DISTANCES ON THE EDGE OF A PIECE OF PAPER. USE AS A SCALE FOR EACH PROBLEM. EACH MARK EQUALS 5.0 mm. (SEE PRACTICE SHEET 17-A FOR EXAMPLE.)

On an A3 sheet, complete the front, top, and right side view for each of the five problems. Remember to locate the front view in the lower left-hand corner of each box. Don't forget to use projection lines and a miter line.

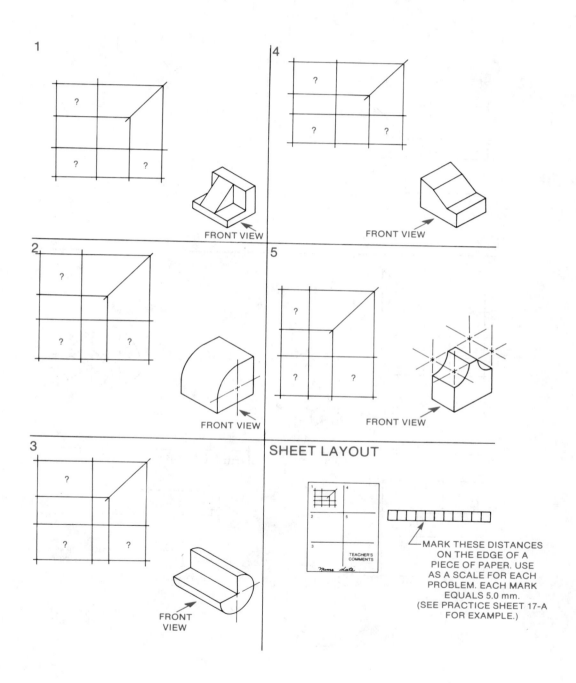

1

2

3

4

5

FRONT VIEW

FRONT VIEW

FRONT VIEW

FRONT VIEW

FRONT VIEW

SHEET LAYOUT

MARK THESE DISTANCES ON THE EDGE OF A PIECE OF PAPER. USE AS A SCALE FOR EACH PROBLEM. EACH MARK EQUALS 5.0 mm. (SEE PRACTICE SHEET 17-A FOR EXAMPLE.)

MULTIVIEW DRAWINGS—SIMPLE OBJECTS 169

On an A3 sheet, complete the views required for each of the five problems. Remember, you need as many views as it takes to represent each feature of the object clearly at least once. Show all projection lines and a miter line.

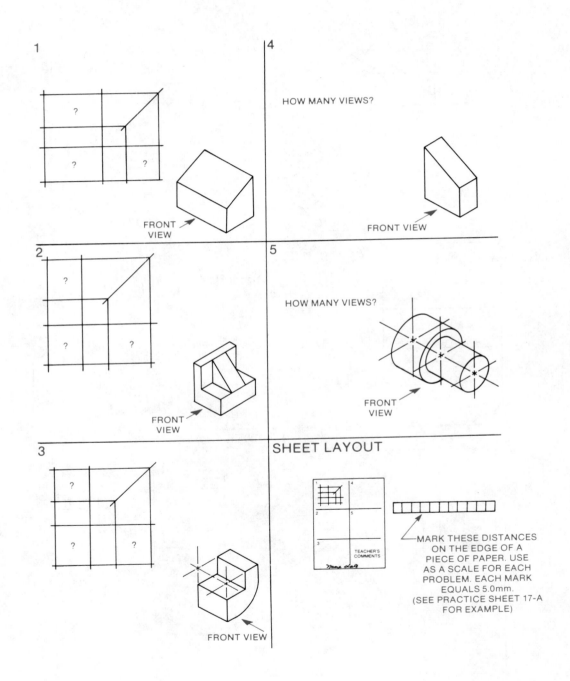

1

?
? ?

FRONT VIEW

4

HOW MANY VIEWS?

FRONT VIEW

2

?
? ?

FRONT VIEW

5

HOW MANY VIEWS?

FRONT VIEW

3

?
? ?

FRONT VIEW

SHEET LAYOUT

TEACHER'S COMMENTS

MARK THESE DISTANCES ON THE EDGE OF A PIECE OF PAPER. USE AS A SCALE FOR EACH PROBLEM. EACH MARK EQUALS 5.0mm. (SEE PRACTICE SHEET 17-A FOR EXAMPLE)

Use an A4 sheet size. Prepare a three-view drawing of the Stop Block. Locate the front view as shown in Figure 17-15. Be sure to use projection lines and a miter line. Locate the title and scale information in the lower right-hand corner.

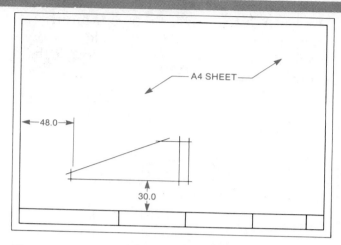

Figure 17-15   *Locating the front view.*

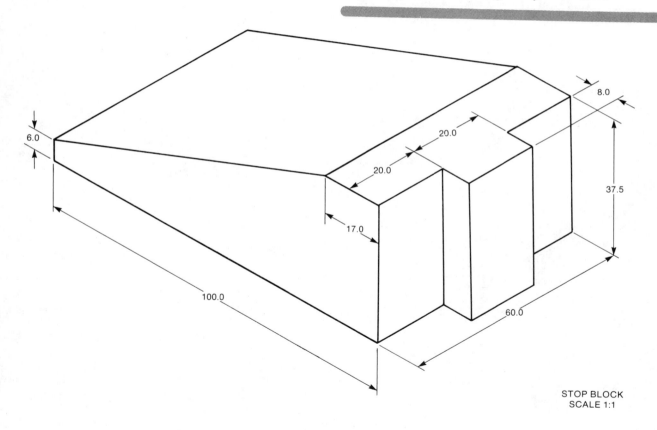

STOP BLOCK
SCALE 1:1

Use an A4 sheet size. Prepare a multi-view drawing of the Hammer Head. How many views are needed? Don't forget the projection lines, title, and scale.

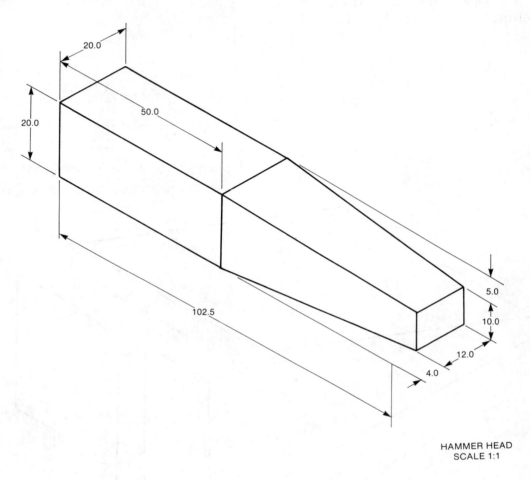

HAMMER HEAD
SCALE 1:1

Using an A4 sheet, prepare a multiview drawing of the Rivet Blank. How many views are needed?

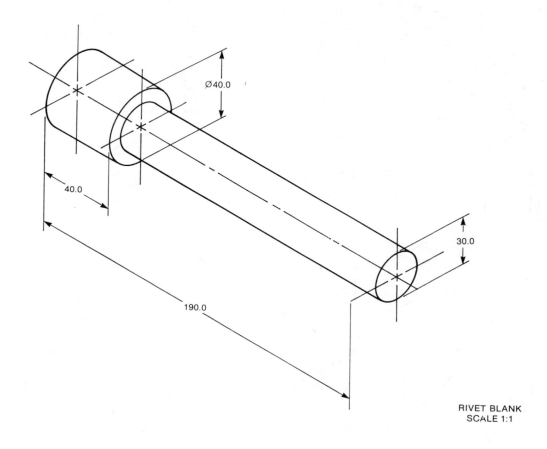

RIVET BLANK
SCALE 1:1

DRAWING PROBLEM 17-4 HANDLE

Using an A3 sheet, prepare a multiview drawing of the Handle. Be sure to use the sheet layout for an A3 sheet. How many views are needed?

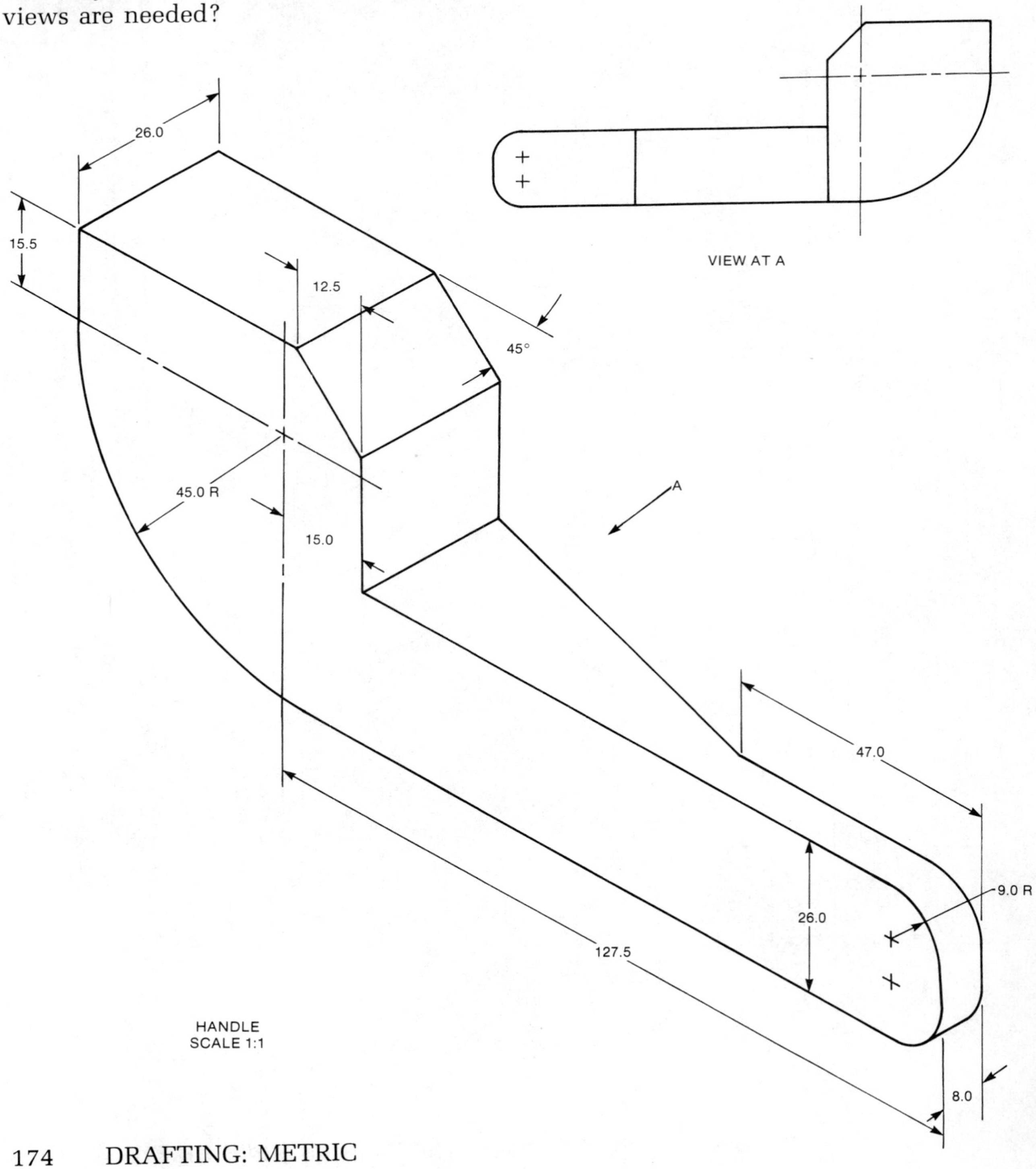

VIEW AT A
26.0
15.5
12.5
45°
45.0 R
15.0
A
47.0
26.0
9.0 R
127.5
8.0

HANDLE
SCALE 1:1

Use an A4 sheet to draw a multiview
drawing of the Stationary Block.

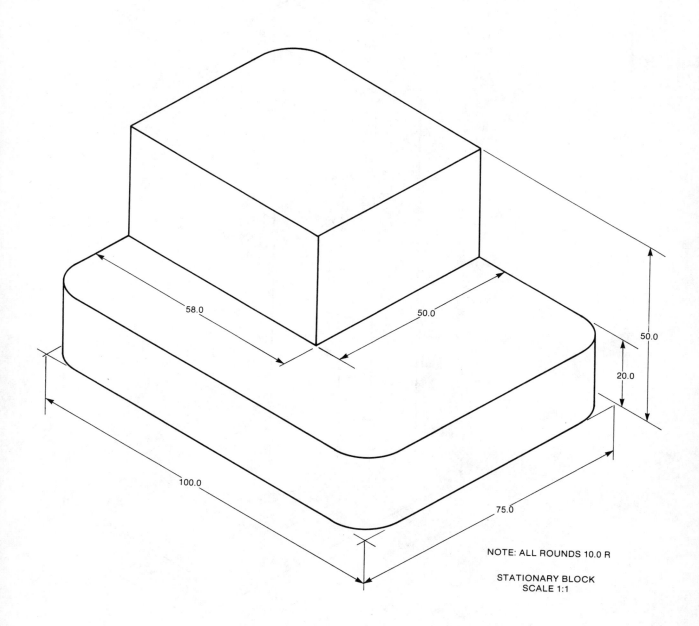

58.0

50.0

100.0

75.0

50.0

20.0

NOTE: ALL ROUNDS 10.0 R

STATIONARY BLOCK
SCALE 1:1

Prepare a multiview drawing of the
Bookend using an A4 sheet.

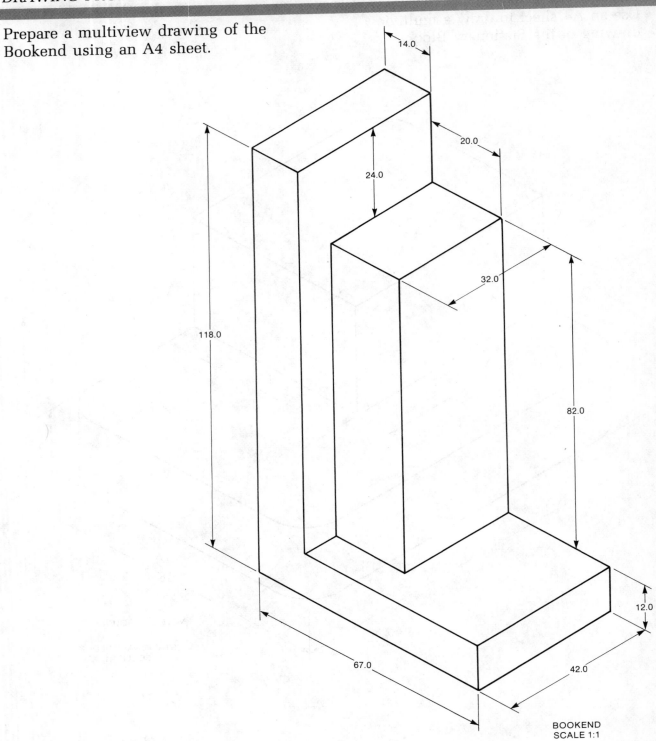

BOOKEND
SCALE 1:1

DRAWING PROBLEM 17-7 TROPHY STAND

Use an A4 sheet to draw a multiview drawing of the Trophy Stand.

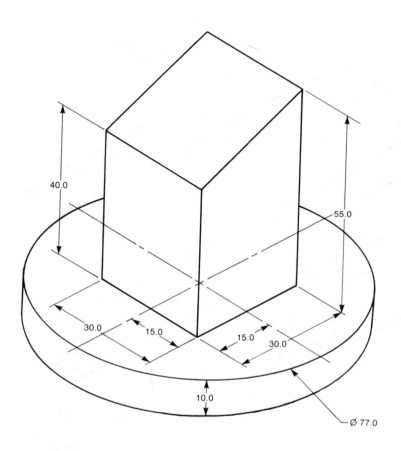

TROPHY STAND
SCALE 1:1

Make a multiview drawing of the Slide
Plate. Use an A4 sheet.

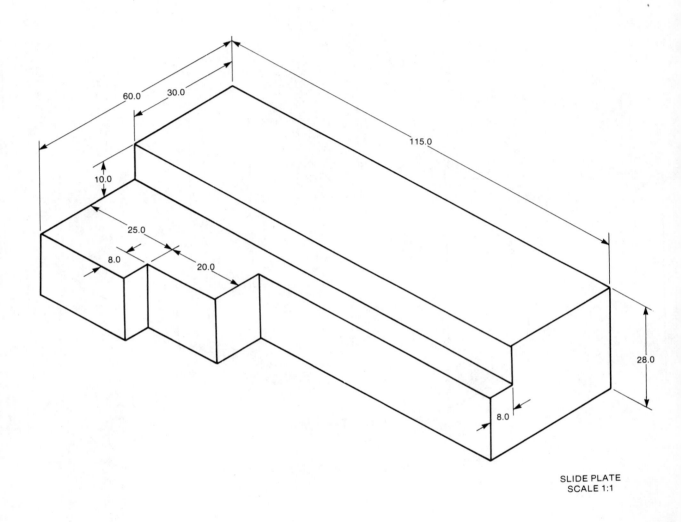

SLIDE PLATE
SCALE 1:1

On an A4 sheet, prepare a multiview
drawing of the Slant Base.

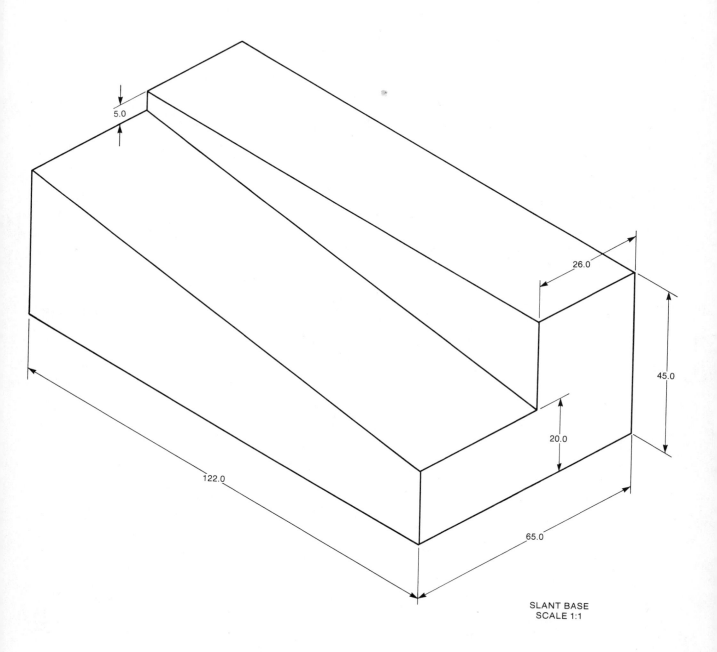

5.0

26.0

45.0

20.0

122.0

65.0

SLANT BASE
SCALE 1:1

Use an A4 sheet. Prepare a multiview
drawing of the Tilt Block.

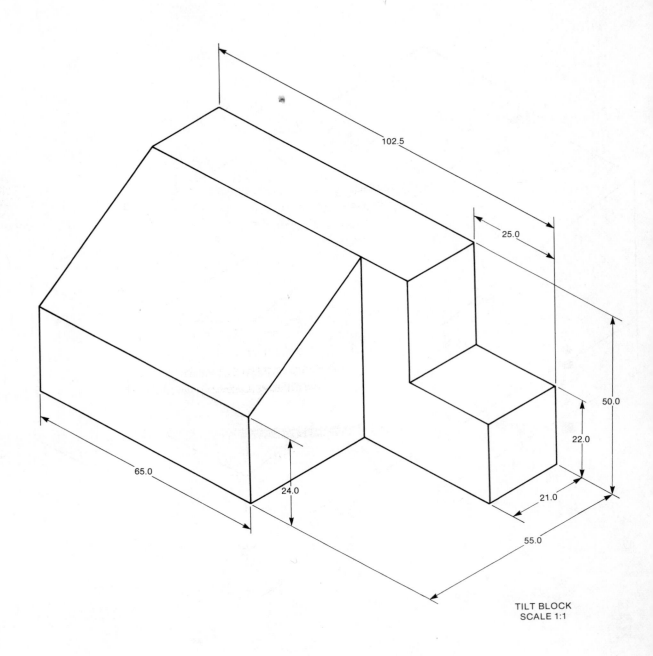

102.5

25.0

50.0

22.0

24.0

65.0

21.0

55.0

TILT BLOCK
SCALE 1:1

- Complicated objects require multiview drawings to represent them.
- The three views of a multiview drawing used most often are the front, top, and right side views.
- Other possible views are the left, bottom, and back side views.
- Distances are transferred from one view to the other by projection lines.
- A miter line is used to change the direction of projection lines from horizontal to vertical or vice versa.
- The number of views used is determined by the number of views required to show each feature of the object clearly at least once.

COMPLEX OBJECTS

MULTIVIEW DRAWINGS * PART 2

This unit shows you how to draw complex objects. You will discover:
- how to draw features that cannot be seen from the outside of an object,
- how to select the best views in drawing an object.

KEY WORDS
Hidden Line: A dashed line used to represent a feature hidden from view.

HIDDEN EDGES
Simple objects present few problems in drawing each feature clearly. As the object becomes more complex, showing each feature becomes more involved. Figure 18-1 shows a complex object that has many hidden edges in the basic views. The dashed lines (*hidden lines*) show the hidden edges.

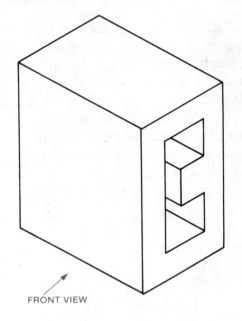

FRONT VIEW

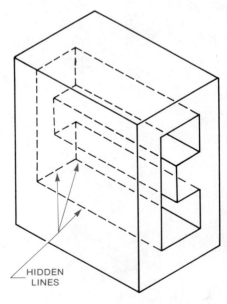

HIDDEN LINES

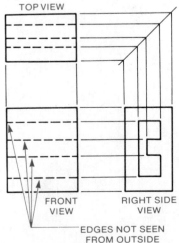

TOP VIEW

FRONT VIEW

RIGHT SIDE VIEW

EDGES NOT SEEN FROM OUTSIDE

Figure 18-1
Hidden lines represent hidden edges.

HIDDEN LINES

A hidden line is drawn as a dashed line. Each dash is approximately 3.0 mm long. The dashes are separated by a space about 1.0 mm wide. The hidden line is as dark as an object line but slightly thinner (Figure 18-2). When drawing hidden lines, the pencil point should be very sharp.

Using Hidden Lines. There are several rules that apply to the use of hidden lines:

1) A hidden line should begin with a dash and end with a dash (Figure 18-3).

2) When a hidden line continues from an object line, it should begin with a space (Figure 18-4).

3) When an object line and a hidden line are to occupy the same space, the object line should be drawn (Figure 18-5).

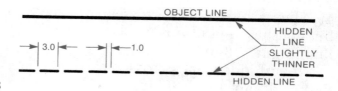

Figure 18-2   *A hidden line.*

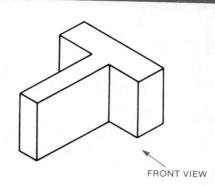

Figure 18-3
A hidden line begins and ends with a dash.

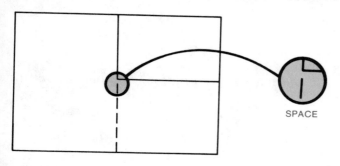

Figure 18-4
A hidden line extending from a solid line.

Figure 18-5
A hidden line and a solid line in the same place.

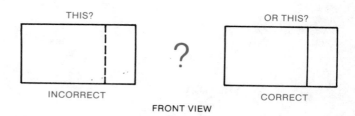

4) When a hidden line crosses an object line, dashes should not touch the object line (Figure 18-6).

5) When hidden lines cross each other, dashes do not touch (Figure 18-7).

6) When hidden lines form a corner, dashes touch (Figure 18-8).

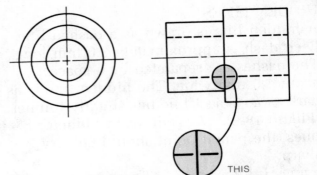

Figure 18-6
A hidden line crossing a solid line.

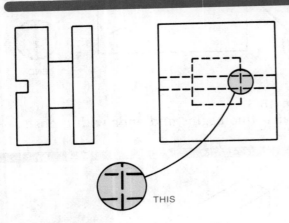

Figure 18-7
A hidden line crossing a hidden line.

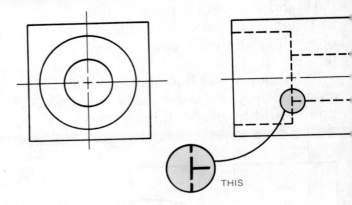

Figure 18-8
A hidden line stopping at another hidden line.

SELECTING THE BEST VIEW WITH HIDDEN EDGES

The best view is always the view with the fewest hidden lines (Figure 18-9).

When selecting the views for an object, the drafter should consider the views that will show each feature clearly (with object lines) at least once (Figure 18-10).

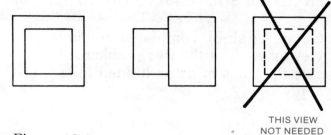

THIS VIEW
NOT NEEDED

Figure 18-9
Select the views with the fewest hidden lines.

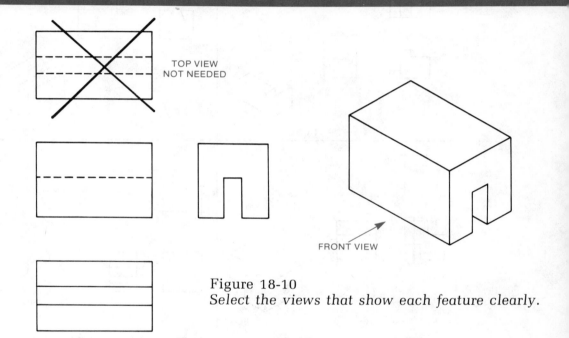

TOP VIEW
NOT NEEDED

FRONT VIEW

Figure 18-10
Select the views that show each feature clearly.

On an A3 sheet, complete the missing view of each of the five problems. Be sure the rules of using hidden lines apply.

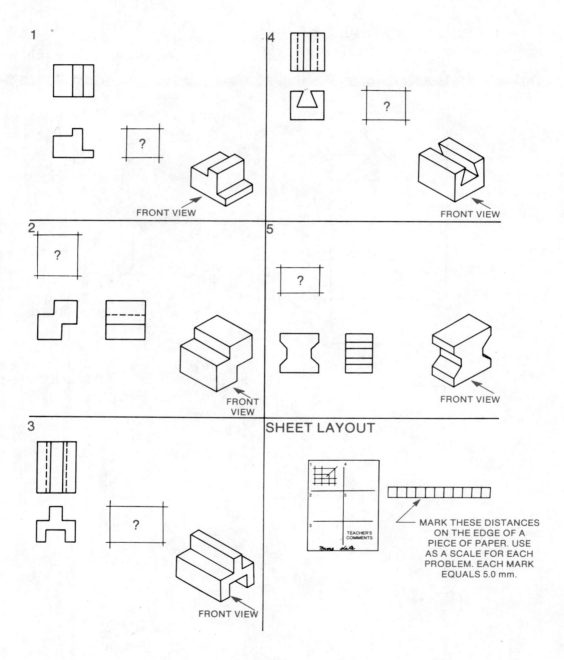

1

? FRONT VIEW

2

? FRONT VIEW

3

? FRONT VIEW

4

? FRONT VIEW

5

? FRONT VIEW

SHEET LAYOUT

TEACHER'S COMMENTS

name date

MARK THESE DISTANCES ON THE EDGE OF A PIECE OF PAPER. USE AS A SCALE FOR EACH PROBLEM. EACH MARK EQUALS 5.0 mm.

On an A3 sheet, complete the missing view of each of the problems. Supply either the front, top, or right side view as indicated. Add all hidden lines that are required.

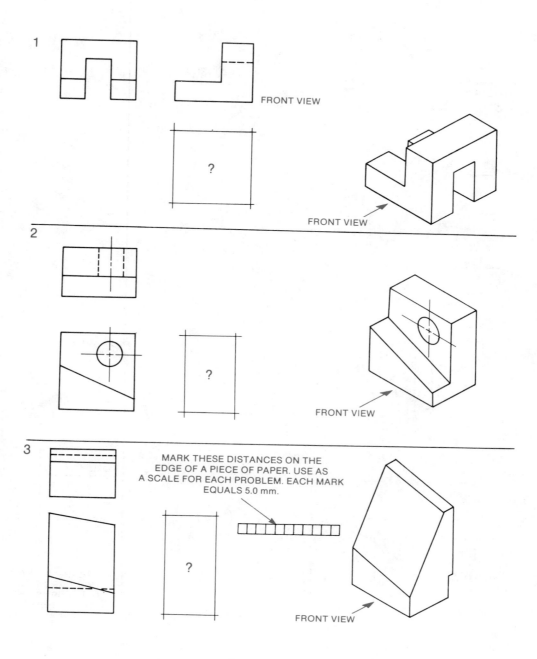

1

FRONT VIEW

?

FRONT VIEW

2

?

FRONT VIEW

3

MARK THESE DISTANCES ON THE EDGE OF A PIECE OF PAPER. USE AS A SCALE FOR EACH PROBLEM. EACH MARK EQUALS 5.0 mm.

?

FRONT VIEW

On an A3 sheet, complete the front, top, and right side view for each of the five problems. The front view is indicated by the words *FRONT VIEW* in the illustration. Add all hidden lines in each view. Use projection lines and a miter line.

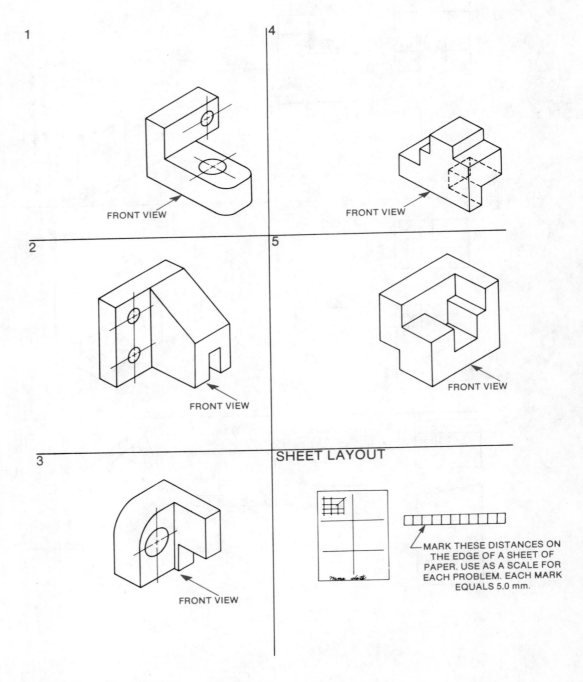

1

FRONT VIEW

4

FRONT VIEW

2

FRONT VIEW

5

FRONT VIEW

3

FRONT VIEW

SHEET LAYOUT

MARK THESE DISTANCES ON THE EDGE OF A SHEET OF PAPER. USE AS A SCALE FOR EACH PROBLEM. EACH MARK EQUALS 5.0 mm.

On an A3 sheet, complete the views required for each of the five problems. Remember, the best view is always the view with the fewest hidden lines.

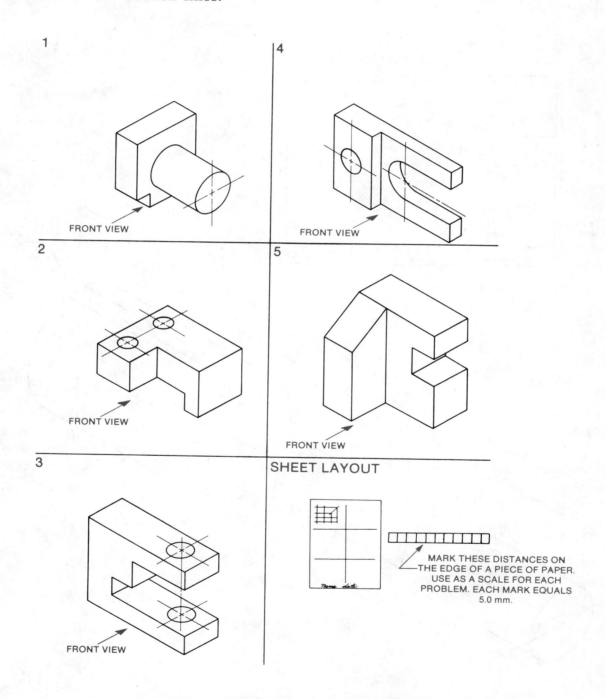

1

FRONT VIEW

2

FRONT VIEW

3

FRONT VIEW

4

FRONT VIEW

5

FRONT VIEW

SHEET LAYOUT

Name date

MARK THESE DISTANCES ON THE EDGE OF A PIECE OF PAPER. USE AS A SCALE FOR EACH PROBLEM. EACH MARK EQUALS 5.0 mm.

DRAWING PROBLEM 18-1 STEPPED SHAFT

Use an A4 sheet size. Prepare a two-view drawing of the Stepped Shaft. Locate the views as shown in the figure. Include the title and scale information.

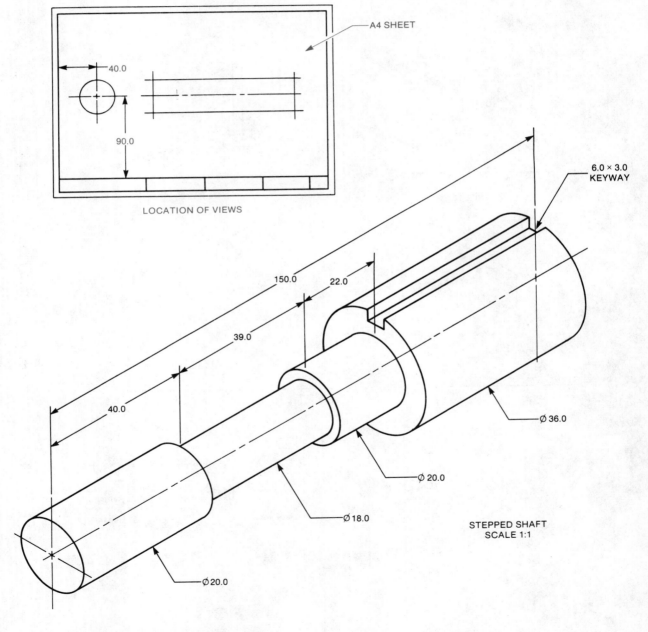

A4 SHEET

40.0

90.0

LOCATION OF VIEWS

6.0 × 3.0
KEYWAY

150.0

22.0

39.0

40.0

Ø 36.0

Ø 20.0

Ø 18.0

STEPPED SHAFT
SCALE 1:1

Ø 20.0

DRAWING PROBLEM 18-2 V-BLOCK

Using an A3 sheet size, prepare a three-view drawing of the V-block. Locate the front view as shown in the figure. Be sure to use projection lines and a miter line. Add the title and scale.

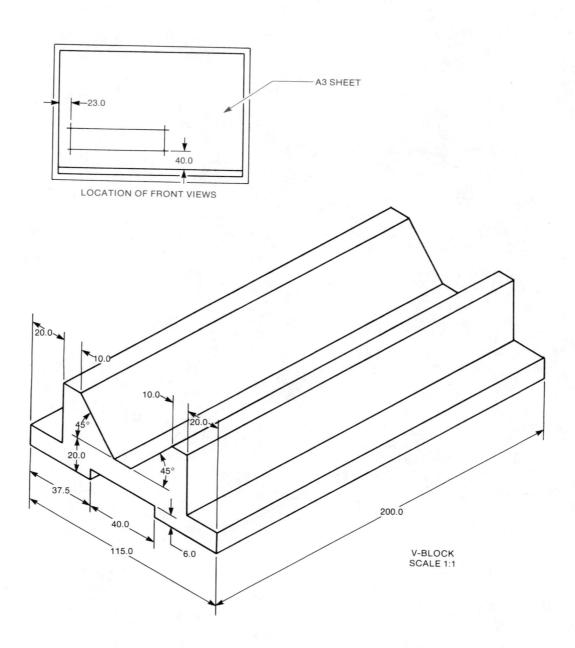

A3 SHEET

23.0

40.0

LOCATION OF FRONT VIEWS

20.0

10.0

10.0

45°

20.0

20.0

45°

37.5

40.0

115.0

6.0

200.0

V-BLOCK
SCALE 1:1

Use an A3 sheet. Prepare a multiview
drawing of the Adjusting Cap. How many
views are required?

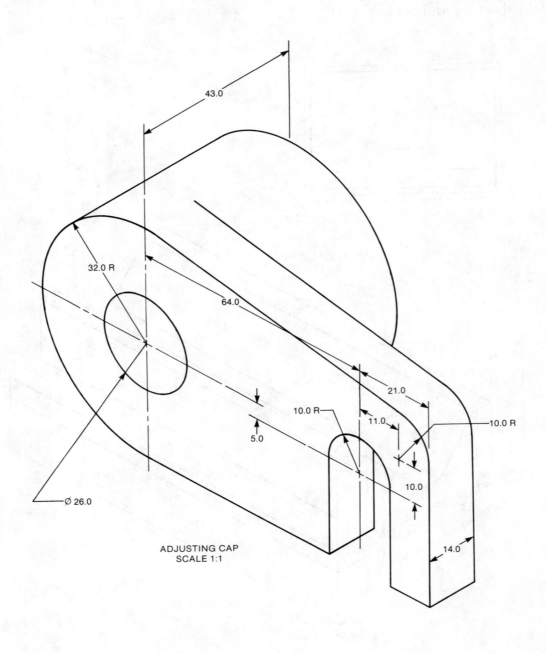

ADJUSTING CAP
SCALE 1:1

DRAWING PROBLEM 18-4 BEARING HOUSING

Use an A3 sheet size. Prepare a multi-view drawing of the Bearing Housing.
How many views are required?

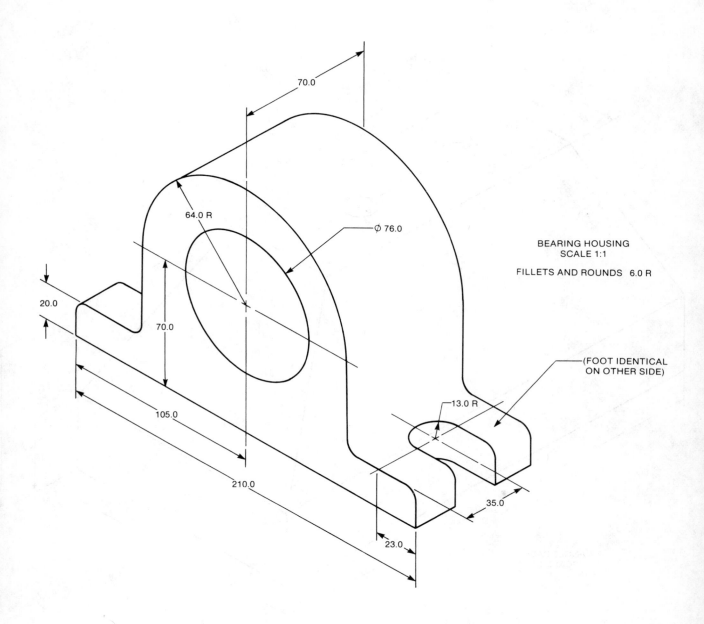

70.0

64.0 R

ⵁ 76.0

BEARING HOUSING
SCALE 1:1

FILLETS AND ROUNDS 6.0 R

20.0

70.0

(FOOT IDENTICAL
ON OTHER SIDE)

13.0 R

105.0

210.0

35.0

23.0

Use an A3 sheet. Prepare a multiview
drawing of the Angle Base.

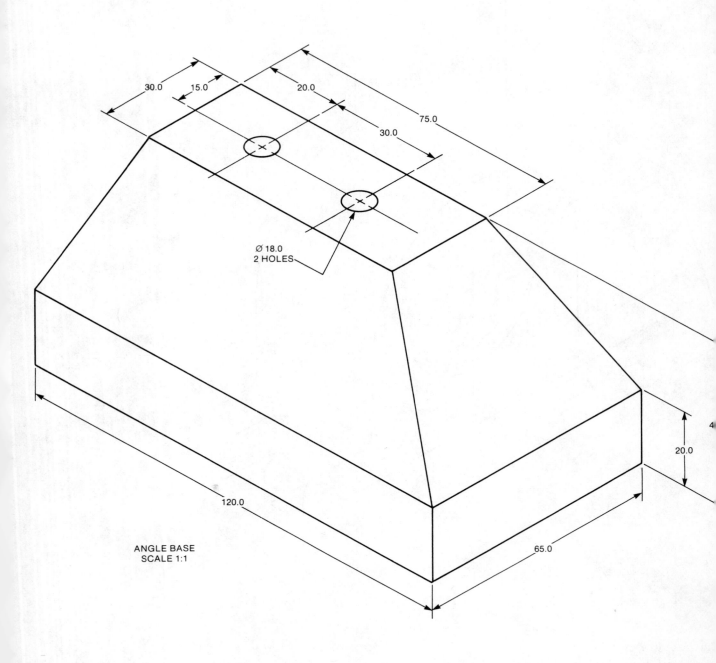

30.0 15.0 20.0 75.0 30.0

Ø 18.0
2 HOLES

120.0

ANGLE BASE
SCALE 1:1

20.0

65.0

On an A3 sheet, prepare a multiview
drawing of the Tube Support Base.

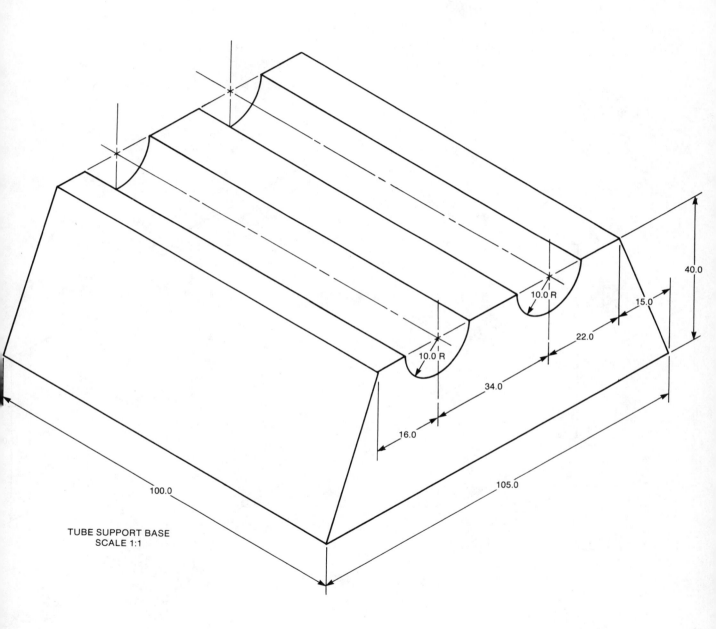

TUBE SUPPORT BASE
SCALE 1:1

Prepare a multiview drawing of the Push
Block. Use an A3 sheet.

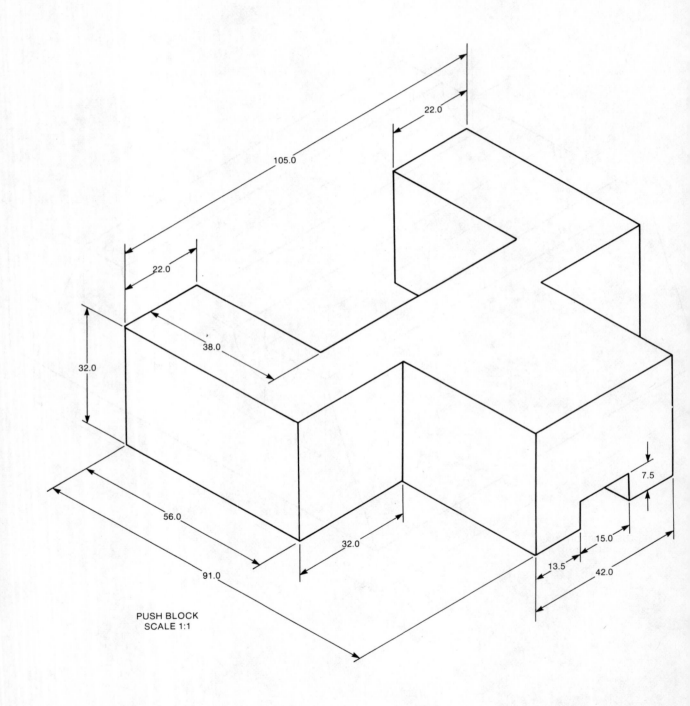

PUSH BLOCK
SCALE 1:1

On an A3 sheet, prepare a multiview
drawing of the Horizontal Stay.

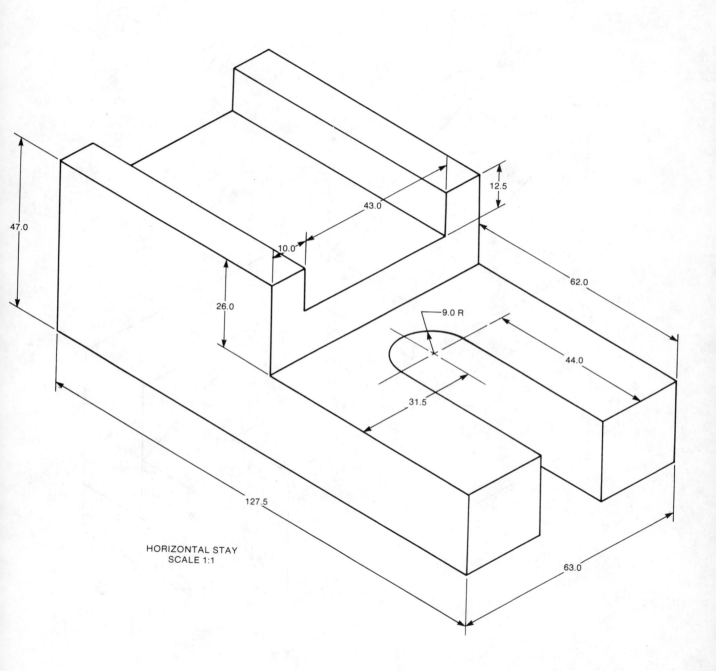

HORIZONTAL STAY
SCALE 1:1

Use an A3 sheet to draw a multiview
drawing of the Feeder Base.

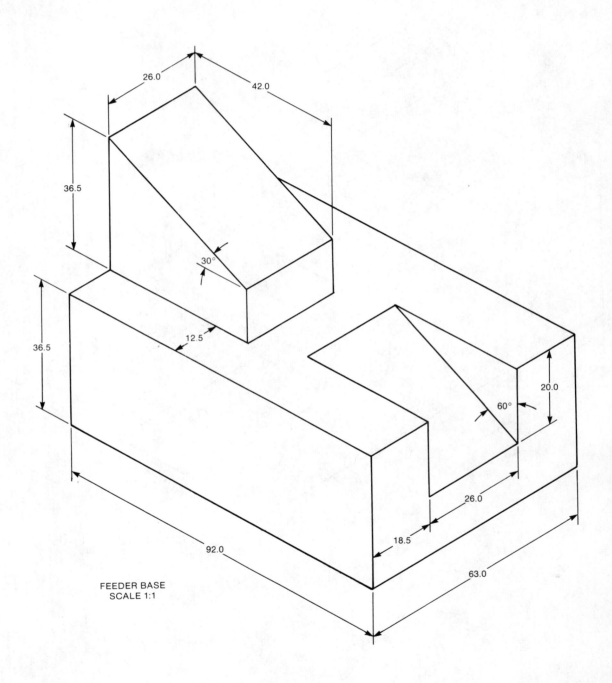

FEEDER BASE
SCALE 1:1

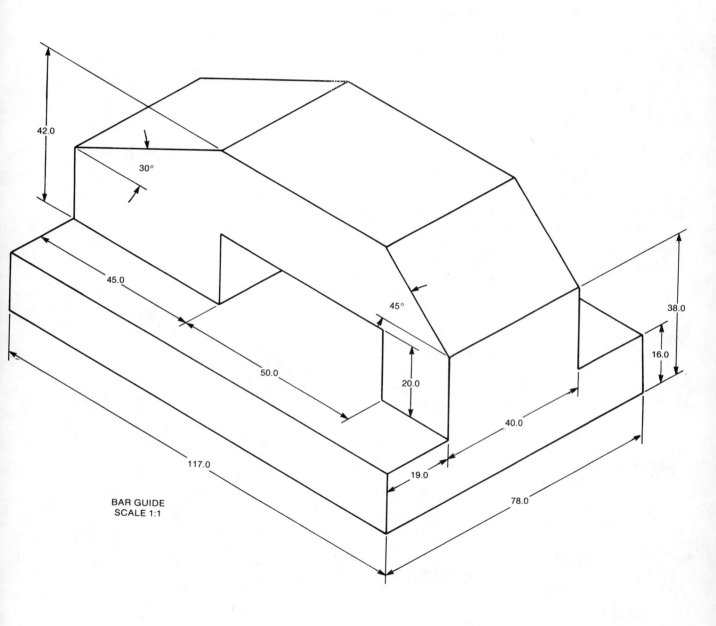

BAR GUIDE
SCALE 1:1

- A hidden line is used to show inside features that cannot be seen from the outside of objects.
- A hidden line is drawn using a series of dashes, each approximately 3.0 mm long.
- The spaces separating the dashes of a hidden line are approximately 1.0 mm long.
- Hidden lines are drawn slightly thinner than object lines.
- When a hidden line crosses an object line, the dashes do not touch the object line.
- When a hidden line crosses another hidden line, the dashes must not touch each other.
- The best view always shows the fewest hidden lines.

DIMENSIONING

This unit shows you how to dimension drawings. You will discover:
- how to describe the sizes of an object,
- how to apply the sizes of an object to a drawing.

KEY WORDS

Dimension: A measurement of an object or portion of an object given on a drawing. The number indicating that measurement (Figure 19-1).

Dimension Line: A fine line that shows the limits of a dimension (Figure 19-1).

Arrowhead: The point of an arrow at the end of a dimension line. It points to the limits of a dimension (Figure 19-2).

Extension Line: A line used to extend a drawn object so that it can be dimensioned (Figure 19-1).

Leader Line: A line from a note to the place the note applies (Figure 19-3).

Note: A statement lettered on a drawing (Figure 19-3).

DIMENSIONS

In order to determine the size of an object, *dimensions* such as length, width, and height are given on a drawing. Including these measurements is called *dimensioning*.

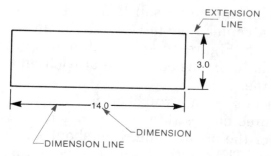

Figure 19-1   *A dimensioned view.*

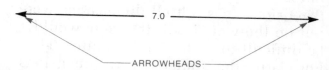

Figure 19-2   *Arrowheads.*

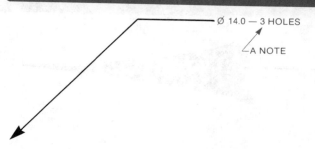

Figure 19-3   *A leader line.*

ALPHABET OF DIMENSION LINES

Just as object lines do, dimension lines have their own purpose and characteristics.

Dimension Lines. A dimension line is used to show the limits of a dimension. The dimension line itself is drawn as thin as a center line and as dark as an object line (Figure 19-4). The dimension line ends in *arrowheads*, one at each end of the line.

Arrowheads. An arrowhead is drawn using three curves (Figure 19-5). The length of the arrow should be about 4.5 mm long. The width should be about 1.5 mm (Figure 19-6). Note the three to one ratio.

Extension Lines. If all dimensions were put on the view itself, the view would be difficult to understand (Figure 19-7). For clarity the drafter places dimensions away from the view (Figure 19-8).

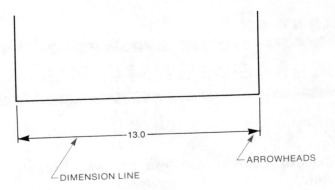

Figure 19-4   *Dimensions.*

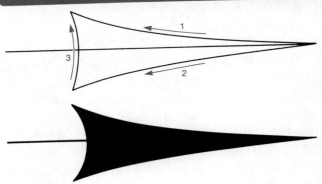

Figure 19-5
Top, *an arrowhead is drawn with curves;* bottom, *an arrowhead completed.*

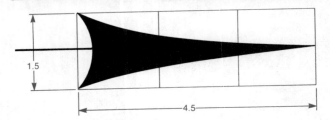

Figure 19-6
The length and width of an arrowhead.

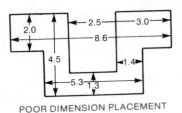

POOR DIMENSION PLACEMENT

Figure 19-7
Incorrect dimensioning.

Figure 19-8   *Proper dimension placement.*

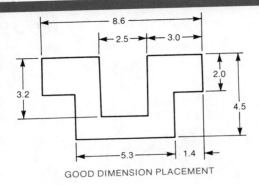

GOOD DIMENSION PLACEMENT

To do this, the drafter must extend the edges of the object (Figure 19-9). These extension lines are drawn dark and thin. The extension line begins about 1.5 mm from the object and extends about 3.0 mm past the arrowhead (Figure 19-10). Extension lines should never cross dimension lines.

Leader Lines. A leader line is used to direct a note to the feature the note is describing (Figure 19-11). Leader lines, like dimension lines, are drawn dark and thin.

A leader line is usually drawn on a 45 degree angle and then bent horizontally (Figure 19-12). If a 45 degree angle is not practical, a 30 or 60 degree angle may be used instead (Figure 19-13).

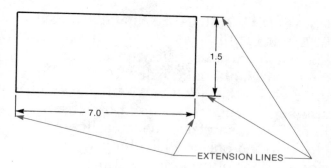

Figure 19-9   Extension lines.

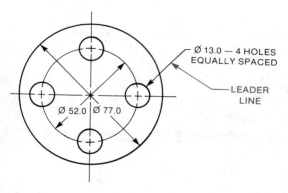

Figure 19-11   A leader line.

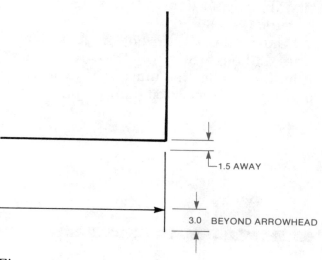

Figure 19-10
Extension lines start away from the view and extend past the arrowhead.

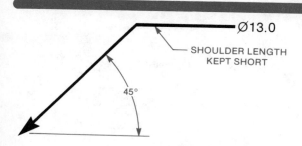

Figure 19-12
A leader line on a 45 degree angle.

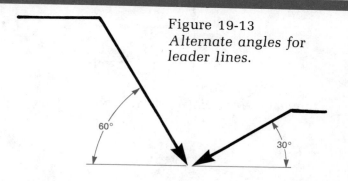

Figure 19-13
Alternate angles for leader lines.

USING DIMENSION, EXTENSION, AND LEADER LINES

Extension and dimension lines should never cross, although extension lines may cross each other (Figure 19-14). To avoid crossing extension lines and dimension lines, locate shorter dimension lines closer to the object. Place longer dimensions farther out (Figure 19-15).

The first dimension line should be at least 10.0 mm away from the object (Figure 19-16). Place additional dimension lines about 7.0 mm away from the first dimension line and from each other (Figure 19-17).

Leader lines should not cross extension lines, although this is not a hard and fast rule. If a leader line must be drawn over an extension line, it should be placed so

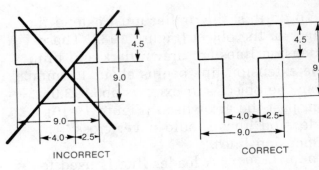

Figure 19-14
Incorrect and correct use of extension lines.

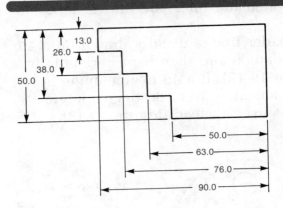

Figure 19-15
Longer dimensions should be placed outside shorter dimensions.

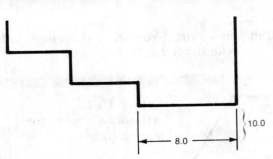

Figure 19-16
Dimension line placed away from view.

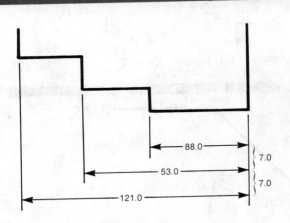

Figure 19-17
Placement of dimension lines from each other.

that it does not interfere with other dimensions (Figure 19-18).

DIMENSION SIZES

Dimensions and notes should be lettered 3.0 mm high (Figure 19-19). Use horizontal guidelines to control the height. Use vertical guidelines to keep the numbers and letters straight (Figure 19-20).

DIMENSION PLACEMENT

A dimension that is poorly placed can be misread. For this reason it is important to locate dimensions correctly.

Rules for Locating Dimensions

1) Show dimensions where they best describe the features of the object (Figure 19-21).

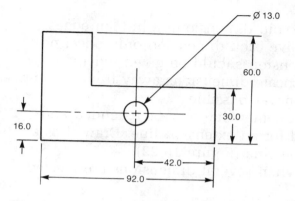

Figure 19-18 *Leader line placement.*

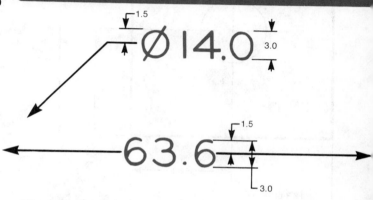

Figure 19-19
Height of dimensions and notes.

Figure 19-20
Guidelines used for dimensions.

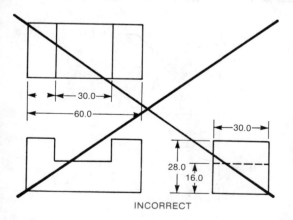

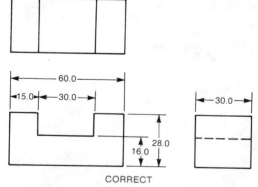

Figure 19-21 *Dimension placement: left, incorrect; right, correct.*

2) Avoid dimensioning hidden edges.
3) Give each dimension only once; no dimension should be given twice.
4) Locate dimensions away from the view whenever possible.
5) A center line may be extended and used for an extension line. Draw it as a center line (Figure 19-22).
6) When several dimensions are located along an edge of the object, stagger the dimensions (Figure 19-23).

DIMENSIONING CIRCLES

The symbol ϕ, which means diameter, should precede the dimension of a diameter (Figure 19-24). The letter R should follow the dimension of a radius. R means radius (Figure 19-25). The

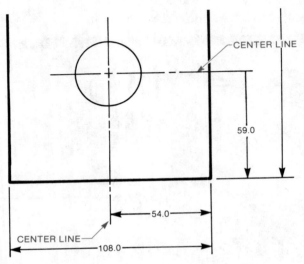

Figure 19-22
Center lines used as extension lines.

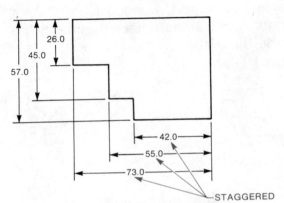

Figure 19-23   *Staggered dimensions.*

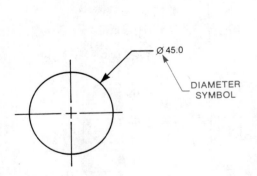

Figure 19-24
The diameter symbol used in dimensioning.

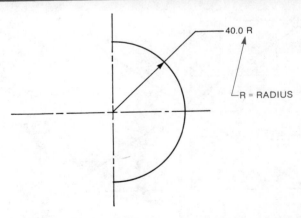

Figure 19-25   *Dimensioning a radius.*

arrowhead of the leader line should touch the edge of the circle (Figure 19-26).

DIMENSIONING RADII
Figure 19-27 shows the proper ways to dimension curves. Always indicate R for radius after the dimension. Note the *break symbol* in the long dimension line. A break symbol is always used if the total dimension line cannot be shown.

DIMENSIONING SMALL DISTANCES
Figure 19-28 shows techniques used to dimension small distances. The size of the space determines whether or not the dimension is placed inside the extension lines.

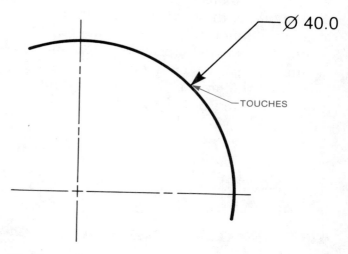

Figure 19-26
The leader line arrowhead touches the circle.

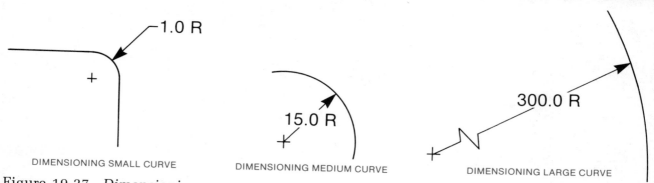

Figure 19-27   Dimensioning curves.

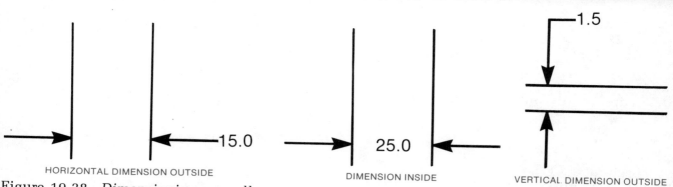

Figure 19-28   Dimensioning a small space.

DIMENSIONING ANGLES

An angular dimension line is curved. The center of the curve is at the vertex of the angle (Figure 19-29). Figure 19-30 shows techniques used for dimensioning angles. Note that the size of the angle determines the positioning of the dimension and arrowheads.

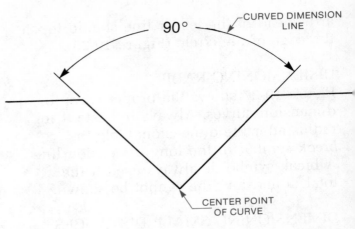

Figure 19-29 *A typical angle dimensioned.*

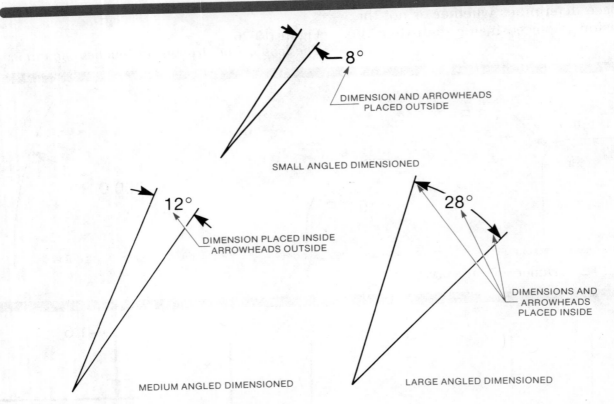

Figure 19-30 *Different sized angles.*

USING NOTES TO DIMENSION

When lettered information or notes are used, several dimensions can be eliminated (Figure 19-31). Write notes in the shortest form possible. Notes that are too "wordy" can cause confusion. Figure 19-32 shows some common notes used on drawings.

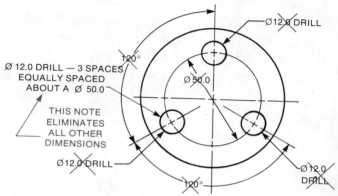

Figure 19-31   Notes eliminate dimensions.

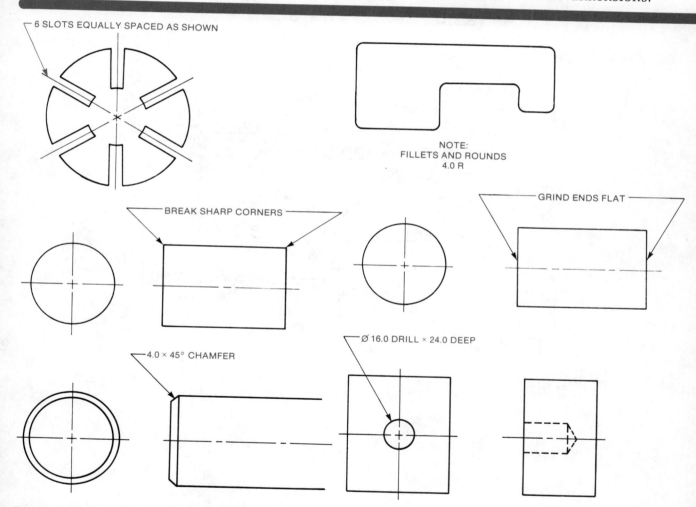

Figure 19-32   Typical notes.

On an A4 sheet, duplicate the dimensioning forms shown. Be sure to review the construction of an arrowhead before you start.

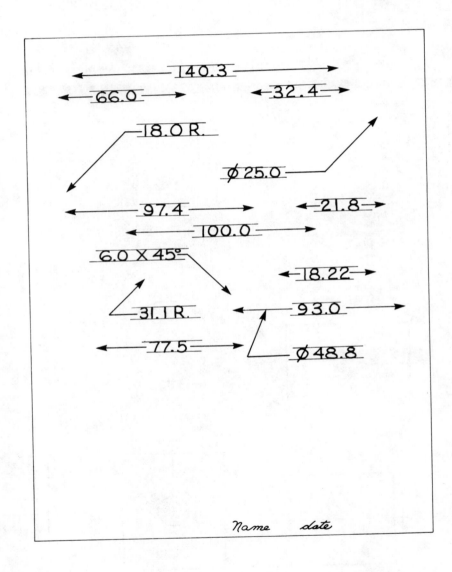

140.3

66.0

32.4

18.0 R.

⌀ 25.0

97.4

21.8

100.0

6.0 X 45°

18.22

31.1 R.

93.0

77.5

⌀ 48.8

Name   date

On an A4 sheet, letter each of the
common notes found on drawings shown
here. Use 3.0 mm high lettering. Be sure
you use light guidelines. Review Unit 12
on lettering before you start.

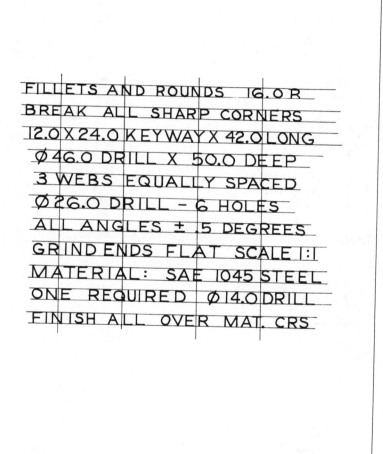

FILLETS AND ROUNDS 16.0 R

BREAK ALL SHARP CORNERS

12.0 X 24.0 KEYWAY X 42.0 LONG

Ø 46.0 DRILL X 50.0 DEEP

3 WEBS EQUALLY SPACED

Ø 26.0 DRILL — 6 HOLES

ALL ANGLES ± 5 DEGREES

GRIND ENDS FLAT SCALE 1:1

MATERIAL: SAE 1045 STEEL

ONE REQUIRED Ø 14.0 DRILL

FINISH ALL OVER MAT. CRS

Name date

On an A3 sheet, sketch the views shown.
Include the dimensions needed for the
figure. Use notes when possible. Place the
dimensions where they best describe the
shape.

1

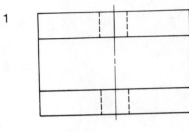

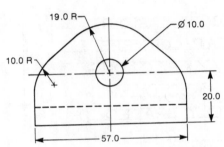

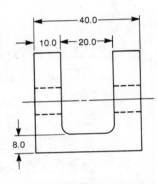

2

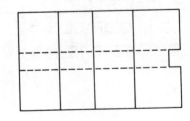

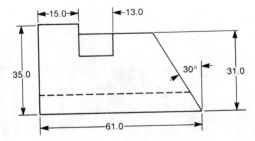

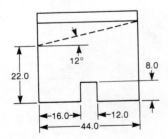

DRAWING PROBLEM 19-1 CYLINDER BRACKET

Use an A3 sheet. Prepare a multiview drawing of the Cylinder Bracket. Dimension the drawing completely. Add all notes, title, and scale information.

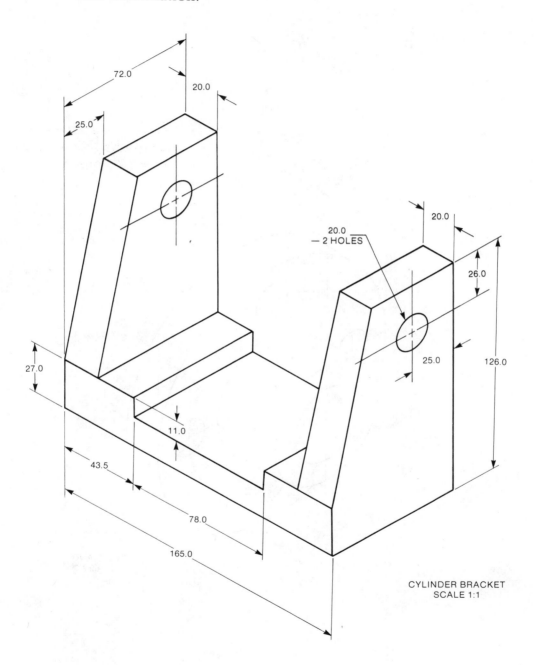

CYLINDER BRACKET
SCALE 1:1

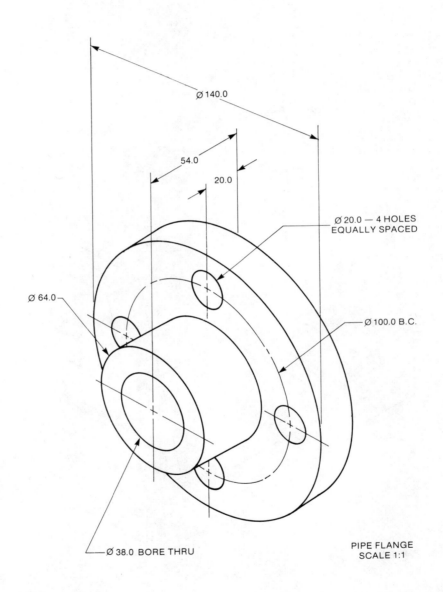

Ø 140.0

54.0

20.0

Ø 20.0 — 4 HOLES
EQUALLY SPACED

Ø 64.0

Ø 100.0 B.C.

Ø 38.0 BORE THRU

PIPE FLANGE
SCALE 1:1

DRAWING PROBLEM 19-3 LINK ARM

Using an A3 size sheet, prepare a multi-view drawing of the Link Arm. Add all dimensions and notes that are needed. Be sure to include the title and scale.

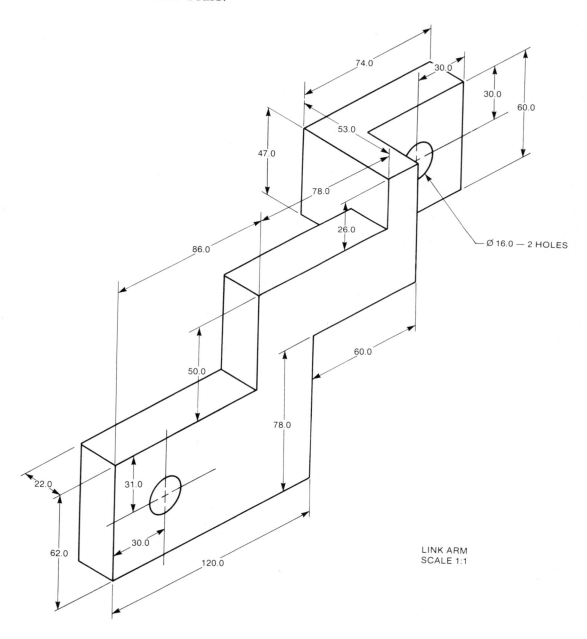

Ø 16.0 — 2 HOLES

LINK ARM
SCALE 1:1

DRAWING PROBLEM 19-4 CRANK HANDLE

Use an A4 sheet. Prepare a multiview drawing of the Crank Handle. Add all dimensions and notes. Add the title and scale information.

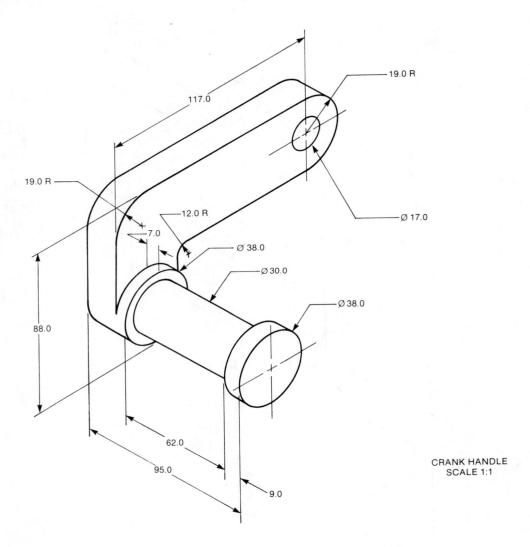

19.0 R

117.0

Ø 17.0

19.0 R

12.0 R

7.0

Ø 38.0

Ø 30.0

Ø 38.0

88.0

62.0

95.0

9.0

CRANK HANDLE
SCALE 1:1

- Dimensions and notes complete the description of an object on a drawing.
- The dimension line indicates the limits of a dimension.
- Arrowheads indicate the limits of a dimension line.
- Extension lines extend the edges of an object for the dimensions.
- Leader lines direct a note to the feature the note describes.
- Extension lines should not cross dimension lines.
- Dimensions and notes should be lettered 3.0 mm high.
- Dimensions should be placed where they best describe the features of an object.
- Dimensions for a diameter should be preceded by the symbol ϕ.
- Dimensions for a radius should be followed by the letter R.
- Notes should be used to eliminate some dimensions.

SCALE DRAWINGS

This unit shows you how to draw large and small objects. You will discover:
- how to draw large objects to a reduced size,
- how to draw small objects to an enlarged size.

KEY WORDS
Proportion: A size or distance in comparison to another.

Proportion Scale: A scale for measuring distances for drawings. Marks on a proportion scale indicate the reduced size in proportion to the full or actual size (Figure 20-1).

SCALE DRAWINGS
When an object is too large to be drawn on a drawing sheet or too small to be seen clearly, the drawing must be reduced or enlarged in proportion to the object. For example, houses are too large to be drawn full size. Wrist watch parts, on the other hand, are too small to be seen clearly when they are drawn full size. Drafting scales can be used to proportionately enlarge or reduce an object on a drawing.

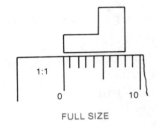

FULL SIZE

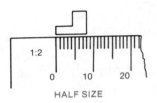

HALF SIZE

Figure 20-1
A full scale and a proportion scale.

PROPORTION SCALES

Most drafting scales are equipped with *proportion scales*. Each of these scales aids the drafter in reducing an object on a drawing in a different proportion. Numbers at the outside edge of the scale indicate the reduction (Figure 20-2). The proportion scales most commonly used for reduction on mechanical drawings are:

Proportion	Means
1:2 (half size)	1 mm on the drawing equals 2 mm actual size
1:5 (one-fifth size)	1 mm on the drawing equals 5 mm actual size
1:10 (one-tenth size)	1 mm on the drawing equals 10 mm actual size
1:20 (one-twentieth size)	1 mm on the drawing equals 20 mm actual size

Using a Proportion Scale. As an example, a 1:2 proportion scale is used to reduce the size of an object to half its actual size on the drawing. The distances between marks on this proportion scale have been reduced by one half (Figure 20-3). Measure the required distances as you would with a full scale (1:1). The final size will then be one half.

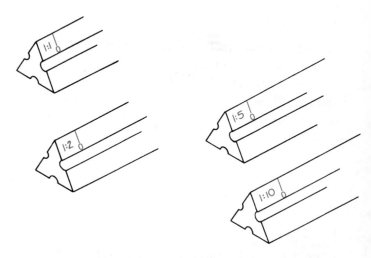

Figure 20-2   *Proportion scales.*

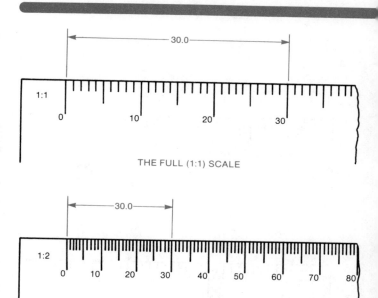

Figure 20-3
The 1:2 scale compared to the 1:1 scale.

Other Proportion Scales. Each of the other proportion scales is used in the same way. In each case the distances between marks have been proportionately reduced. Figure 20-4 shows 1:5 proportion; Figure 20-5 shows 1:10 proportion; Figure 20-6 shows 1:20 proportion.

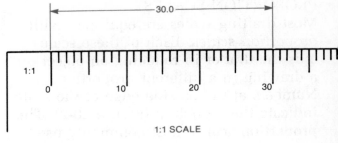

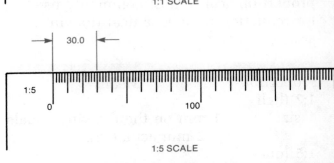

Figure 20-4
The 1:5 scale compared to the 1:1 scale.

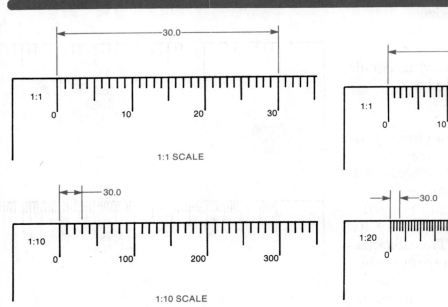

Figure 20-5
The 1:10 scale compared to the 1:1 scale.

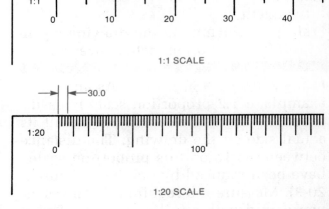

Figure 20-6
The 1:20 scale compared to the 1:1 scale.

PRACTICE SHEET 20-A PROPORTION SCALES

On an A4 sheet, draw four 100.0 mm
lines using the proportion scales
indicated in the figure. Below each line,
letter the proportion scale used.

SCALE 1:1

SCALE 1:2

SCALE 1:5

SCALE 1:20

ADDITIONAL PROBLEMS

Using the scales 1:1, 1:2, 1:5, and 1:20,
draw each of the following lines: 150.0,
75.0, 90.0, 120.0.

THE ENLARGEMENT SCALES

To enlarge an object on a drawing, you do not need special scales. Instead the 1:1 scale is used, but it must be read differently.

Drawing to an Enlarged Scale. To draw an object twice as large as it actually is (2:1 scale), use the full (1:1) scale and read each mark as if it represented 0.5 mm instead of 1.0 mm; for example, read the 10.0 mm mark as 5.0 mm (Figure 20-7). In this way the scale enlarges the drawing proportionately (Figure 20-8).

Drawing Objects Larger Than 2:1 Scale. Drawing objects larger than double (2:1 scale) is rarely needed for mechanical drawings. To draw objects larger than double scale, however, you can use the full (1:1) scale and read it as if each mark represented a smaller distance. Figure 20-9 shows how to read a 10:1 scale.

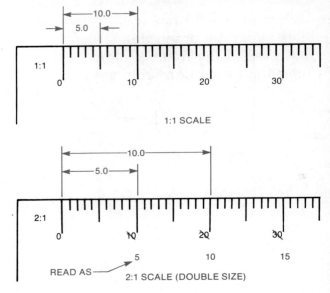

Figure 20-7   *Measuring to a scale of 2:1.*

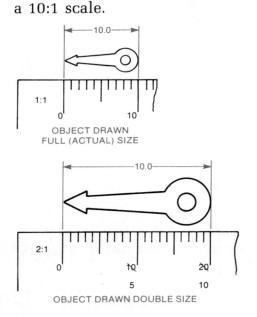

Figure 20-8
The 2:1 scale measures distances to twice the actual size.

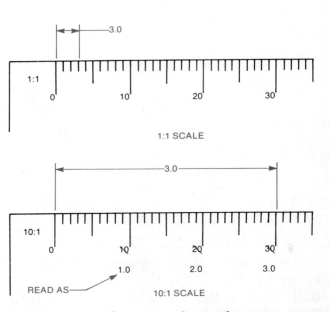

Figure 20-9   *The 1:1 scale used as 10:1.*

PRACTICE SHEET 20-B PROPORTION SCALES

On an A4 sheet, draw each of the five
lines to the double (2:1) scale. Dimension
each length for each line as shown in the
figure.

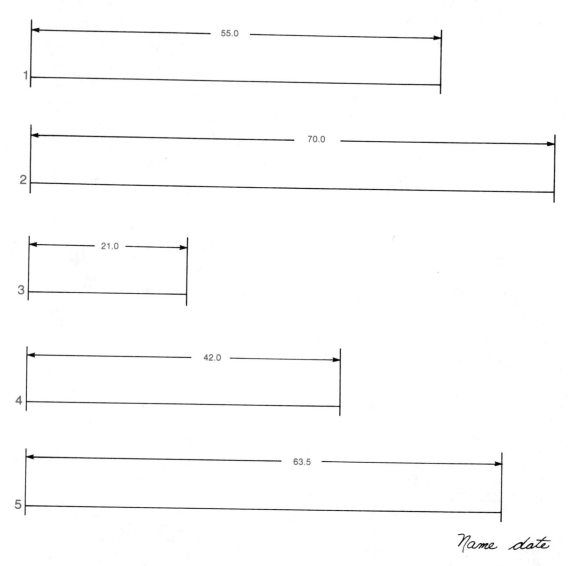

ADDITIONAL PROBLEMS

Draw each of the following lines using
the 2:1 scale: 62.0, 90.0, 20.0, 76.0, 14.0.

DRAWING PROBLEM 20-1 LIFTING FORK

Use an A3 sheet. Prepare a multiview drawing of the Lifting Fork. Use the proportion scale 1:5. Dimension the drawing completely. Add all notes and the title. Add the scale information.

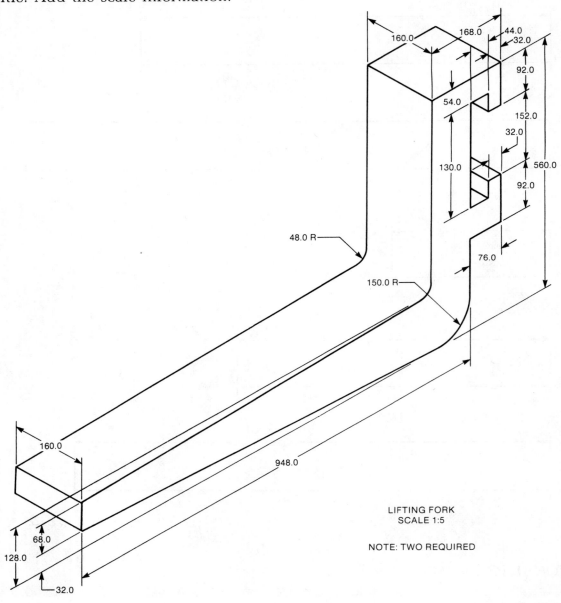

LIFTING FORK
SCALE 1:5

NOTE: TWO REQUIRED

DRAWING PROBLEM 20-2 TABLE BASE

Using an A4 sheet, prepare a multiview
drawing of the Table Base. Use the
proportion scale of 1:10. Dimension fully.
Add all notes, title, and scale
information.

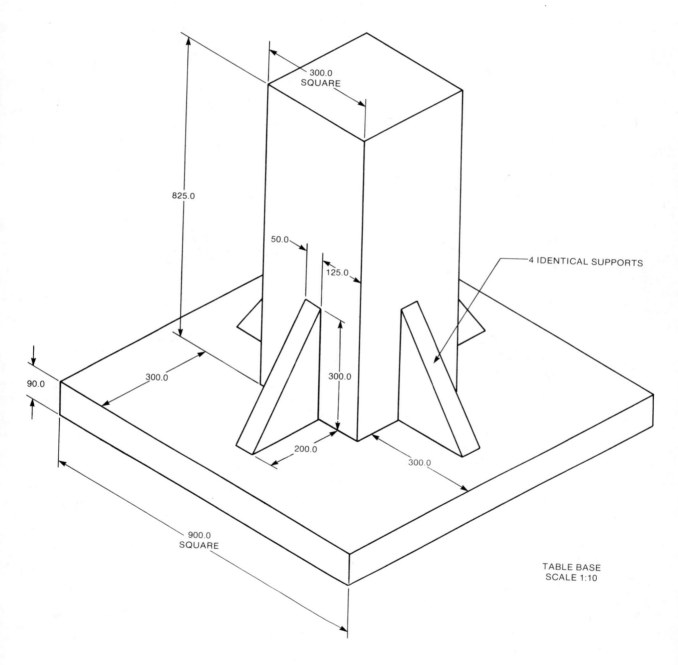

300.0
SQUARE

825.0

50.0

125.0

4 IDENTICAL SUPPORTS

300.0

300.0

90.0

200.0

300.0

900.0
SQUARE

TABLE BASE
SCALE 1:10

On an A4 sheet, prepare a multiview drawing of the Handle. Use the scale 2:1 (double size). Dimension fully. Add all notes, title, and scale information.

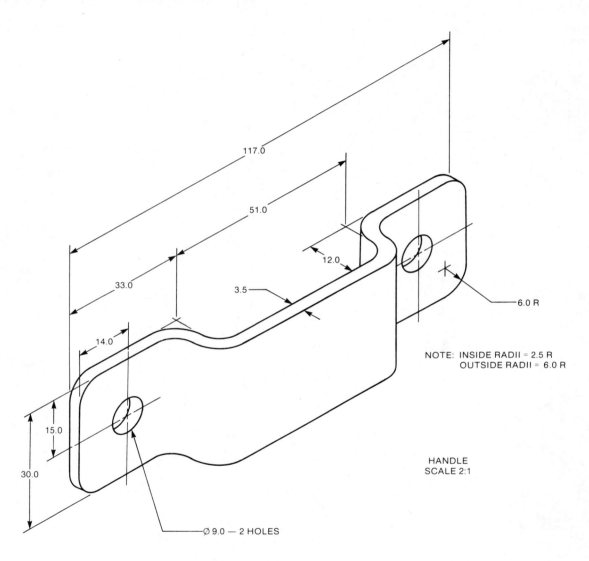

117.0

51.0

33.0

12.0

3.5

14.0

6.0 R

NOTE: INSIDE RADII = 2.5 R
OUTSIDE RADII = 6.0 R

15.0

30.0

HANDLE
SCALE 2:1

Ø 9.0 — 2 HOLES

- When an object is too large to be drawn on a drawing sheet, a proportion scale for reduction is used.
- When an object is too small to be seen clearly, a proportion scale for enlargement is used.
- The most common scales for reduction are 1:2, 1:5, 1:10, and 1:20.
- The distance between marks on a proportion scale has been reduced to reflect the proportion desired.
- To enlarge an object to twice its size on a drawing, the full scale markings are read as 0.5 mm instead of 1.0 mm.

SECTIONS

This unit will show you how to draw objects with the inside exposed. You will discover:
- how to draw objects with complicated interiors,
- how to draw objects using sectioned views.

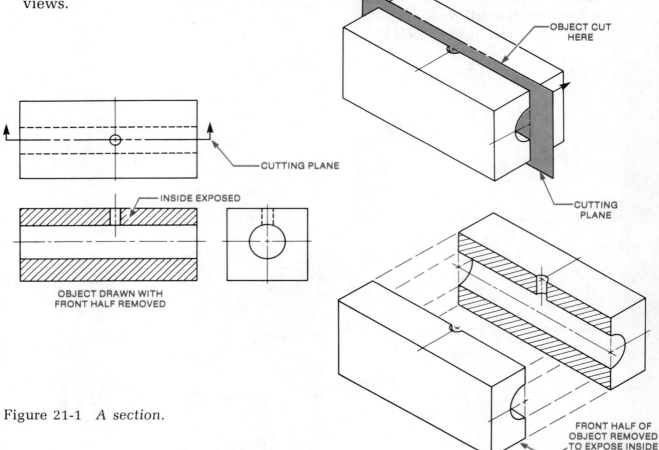

Figure 21-1   *A section.*

KEY WORDS

Section: A drawing of an object with a portion removed for clarity (Figure 21-1).

Section Lining: A series of thin, parallel lines used to indicate the portion of an object which is cut for the purpose of a section (Figure 21-1).

Cutting Plane Line: A line used to indicate where an object has been cut for a section view (Figure 21-1).

SECTION VIEWS

The regular views of a multiview drawing may not show the entire object clearly. Sometimes, the inside of an object cannot be seen easily because of hidden lines (Figure 21-2). To help show the complicated inside of objects, the drafter will draw the object with a portion removed. This is called a *section* drawing (Figure 21-2).

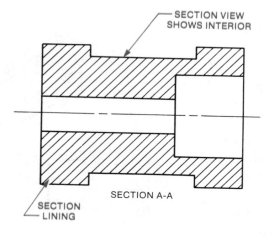
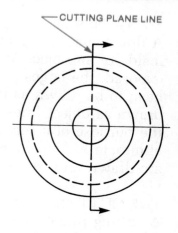

SECTION VIEW
SHOWS INTERIOR

CUTTING PLANE LINE

SECTION A-A

SECTION LINING

Figure 21-2
An object shown sectioned.

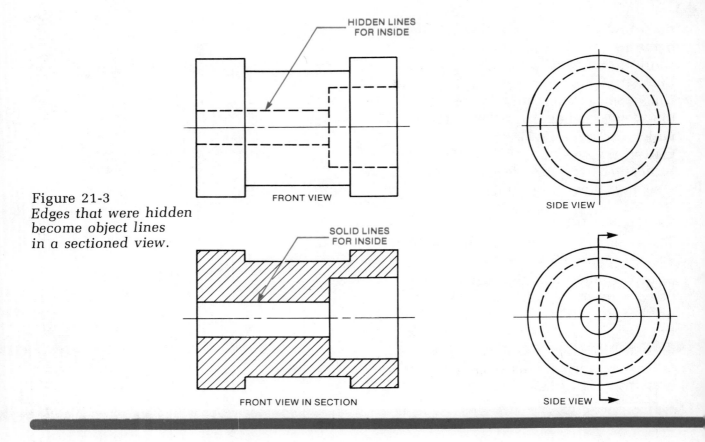

Figure 21-3
*Edges that were hidden
become object lines
in a sectioned view.*

HIDDEN LINES
FOR INSIDE

FRONT VIEW

SIDE VIEW

SOLID LINES
FOR INSIDE

FRONT VIEW IN SECTION

SIDE VIEW

SECTIONING

A drafter uses sectioning to show the
inside of an object. When the inside is
exposed, the edges or features that would
have to be drawn as hidden lines are
drawn as object lines (Figure 21-3). Since
the purpose of a sectioned view is to
eliminate hidden lines, no hidden lines
are drawn on a sectioned view.

THE CUTTING PLANE LINE

A *cutting plane line* indicates the point at
which an object is being sectioned
(Figure 21-4). A cutting plane line is a
series of long and short dashes drawn

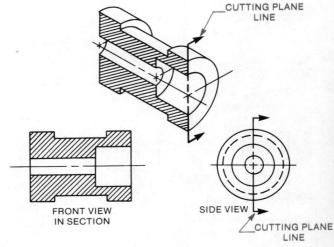

CUTTING PLANE
LINE

FRONT VIEW
IN SECTION

SIDE VIEW

CUTTING PLANE
LINE

Figure 21-4
*The cutting plane shows from where a
section is viewed.*

thick and dark. The long dash is drawn between 15.0 mm and 20.0 mm long. The short dash is about 3.0 mm long. See Figure 21-5.

Arrowheads are placed at each end of the cutting plane line to indicate from which direction the section is being viewed (Figure 21-5).

On drawings with more than one section view, and therefore more than one cutting plane, each cutting plane and section view is identified by letters (Figure 21-6). Place the letters behind the arrows of the cutting plane line. Below the sectioned view, place a subtitle, such as "SECTION A-A" (Figure 21-6).

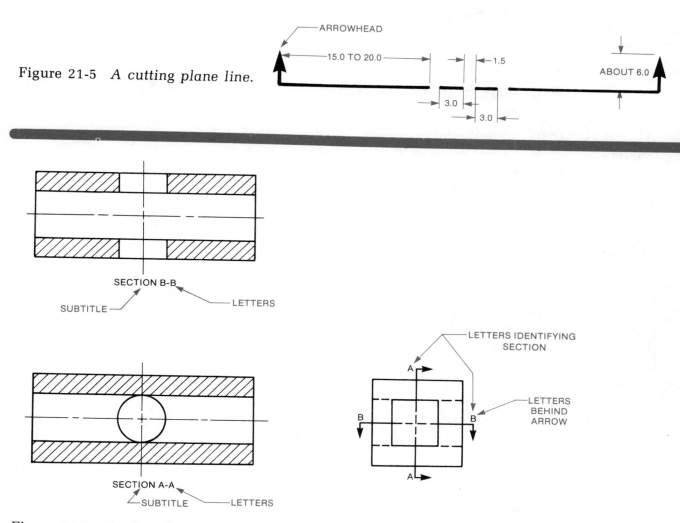

Figure 21-5   *A cutting plane line.*

Figure 21-6   *Cutting planes and sectioned views are identified by letters.*

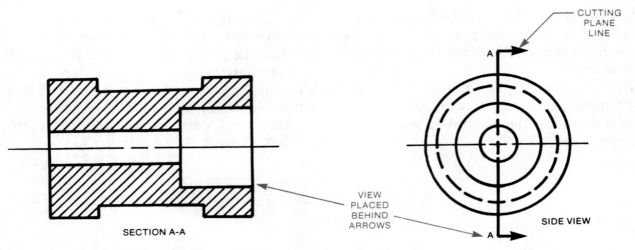

SECTION A-A

VIEW PLACED BEHIND ARROWS

CUTTING PLANE LINE

SIDE VIEW

Figure 21-7 *A full section.*

THE FULL SECTION

When a sectioned view is drawn as if one-half of the object has been removed, the view is called a *full section* (Figure 21-7). Note that the sectioned view is drawn behind the arrows of the cutting plane (Figure 21-7).

THE HALF SECTION

When a sectioned view is drawn as if one-fourth of the object has been removed, the view is called a *half section* (Figure 21-8). In a half section drawing, the line separating the section from the nonsectioned portion remains a center line rather than a cutting plane line. The quarter portion is drawn as any other external view. Note that hidden lines are omitted.

Figure 21-8 *A half section.*

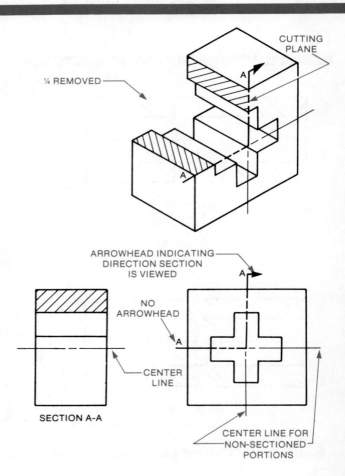

¼ REMOVED

CUTTING PLANE

ARROWHEAD INDICATING DIRECTION SECTION IS VIEWED

NO ARROWHEAD

CENTER LINE

SECTION A-A

CENTER LINE FOR NON-SECTIONED PORTIONS

BROKEN-OUT SECTIONS

When a full section or a half section would expose more than is needed, a *broken-out section* (Figure 21-9) may be used. A broken-out section shows the view with only a portion removed. A jagged line, used to separate the sections, is drawn to give the appearance that the front has been broken away.

REMOVED SECTIONS

A section may be set away from a view. This type of section is called a removed section (Figure 21-10). As with other sections, the cutting plane indicates the point from which the section is being viewed.

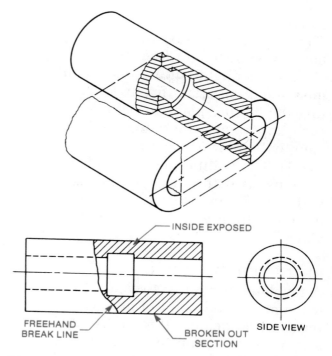

Figure 21-9 *A broken out section.*

Figure 21-10 *Removed sections.*

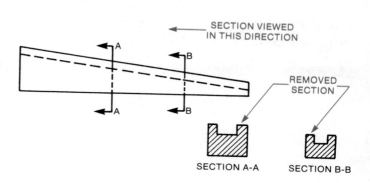

OFFSET SECTIONS

The cutting plane does not necessarily have to be a straight line. With some objects, a straight cutting plane could miss some important features. By offsetting the cutting plane to pass through each important feature, the sectioned view will show an object clearly (Figure 21-11). The cutting plane for offset sections, as with other sections, is drawn with arrowheads at each end to indicate the point from which the section is being viewed.

REVOLVED SECTIONS

A revolved section is a section that is taken at a given point of an object and revolved 90 degrees (Figure 21-12). The section is drawn on the same view at the point it is taken (Figure 21-12). There is no separate section view. Figure 21-13 shows some typical revolved sections.

SECTION LINING

To fill in the area in a sectioned view that is "cut" by the cutting plane *section lining* is used. Section lining is drawn as thin and as dark as a center line.

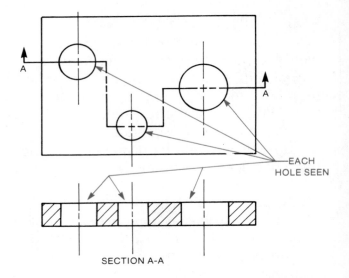

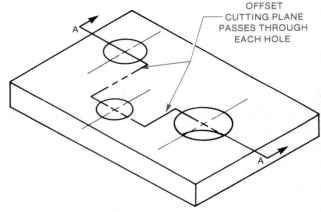

Figure 21-11   *An offset section.*

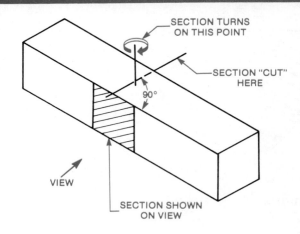

Figure 21-12   *A revolved section.*

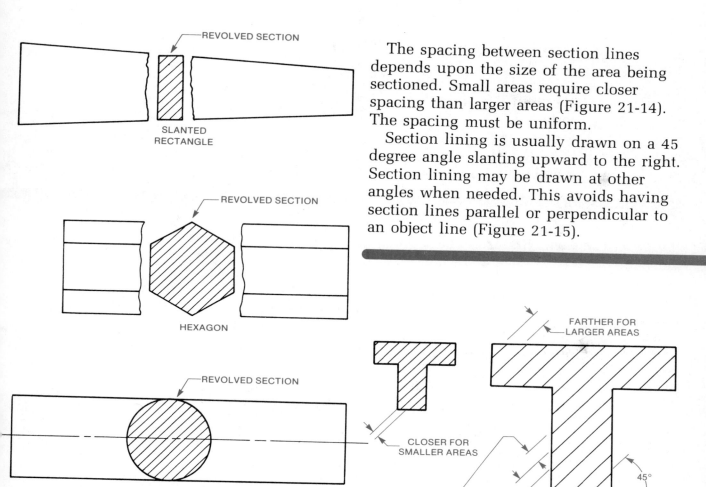

The spacing between section lines depends upon the size of the area being sectioned. Small areas require closer spacing than larger areas (Figure 21-14). The spacing must be uniform.

Section lining is usually drawn on a 45 degree angle slanting upward to the right. Section lining may be drawn at other angles when needed. This avoids having section lines parallel or perpendicular to an object line (Figure 21-15).

Figure 21-13   Typical revolved sections.

Figure 21-14   Section lining.

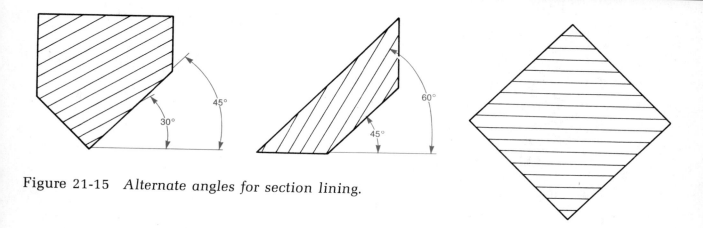

Figure 21-15   Alternate angles for section lining.

RIBS, WEBS, AND SPOKES IN SECTION

Ribs, webs, and spokes are thin portions that are not continuous around an object (Figure 21-16). When an object is being sectioned through a rib, web, or spoke, section lining is omitted (Figure 21-17).

Some objects, when drawn in section, may appear distorted (Figure 21-18).

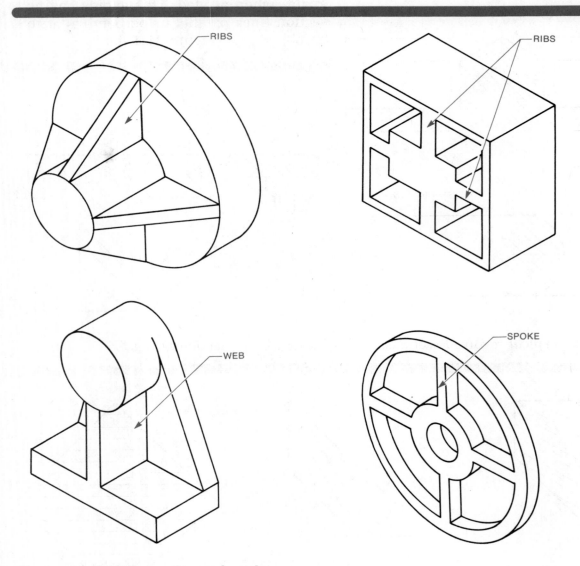

Figure 21-16   *Ribs, webs, and spokes.*

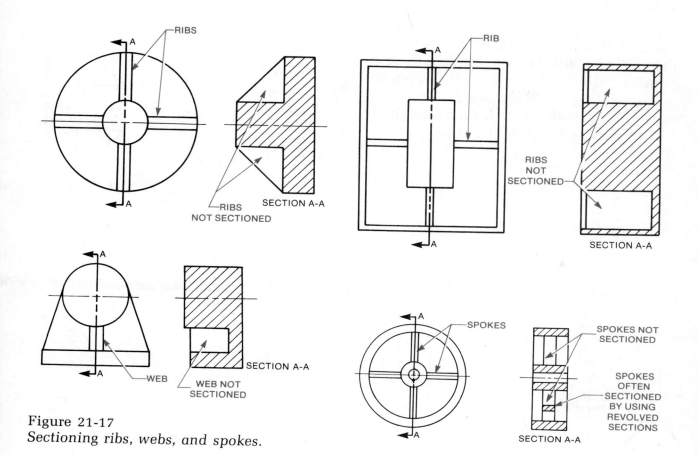

Figure 21-17
Sectioning ribs, webs, and spokes.

Figure 21-18 Rib appears distorted.

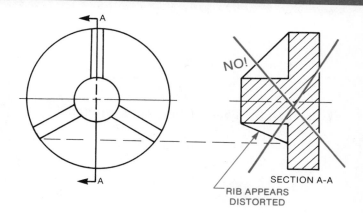

Features that may appear to be away from the cutting plane (ribs, webs, spokes, and holes) should be drawn as if they appear on the cutting plane (Figure 21-19). Rotate the cutting plane from a straight line to pass through the features (Figure 21-20).

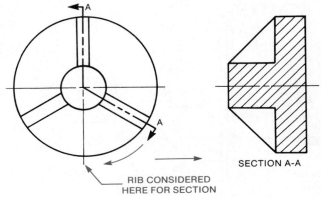

Figure 21-19
Ribs, webs, spokes, and holes drawn to eliminate distortion.

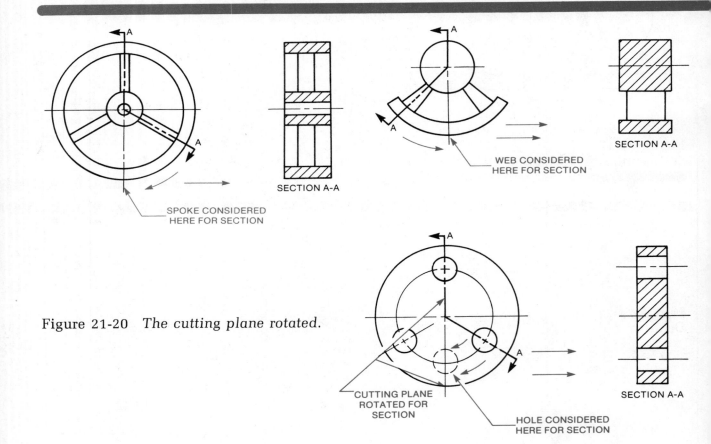

Figure 21-20　*The cutting plane rotated.*

CONVENTIONAL BREAKS

There can be problems in fitting long, thin objects to the sheet (Figure 21-21). If the object is drawn in reduced proportion, the length may fit the sheet, but the thin portion may become too small to be seen (Figure 21-22).

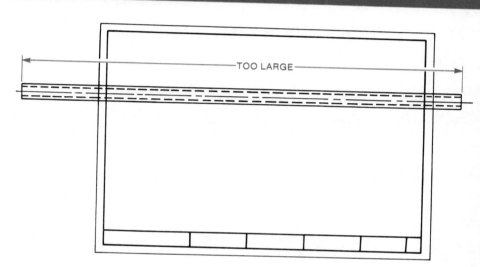

Figure 21-21
An object too large for the drawing sheet.

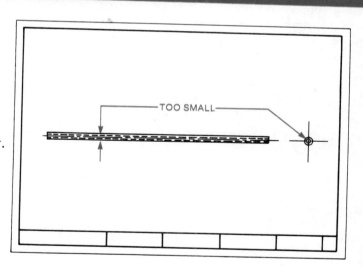

Figure 21-22
Detail of object too small to be seen clearly.

To solve this problem, use conventional breaks (Figure 21-23). Conventional breaks show that the object is much longer than is drawn. Figure 21-24 shows

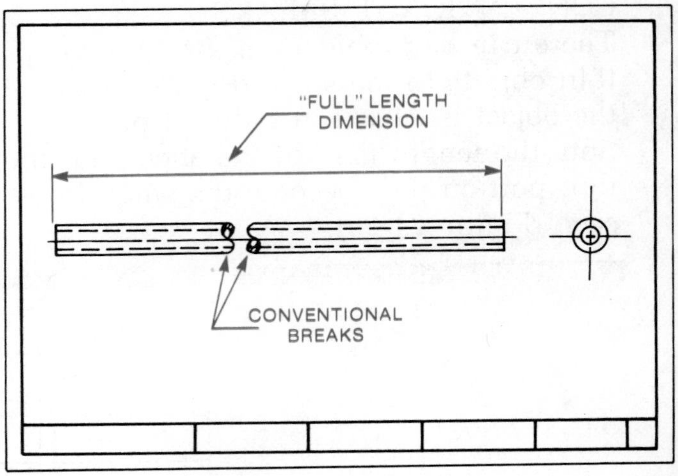

Figure 21-23   *Conventional break.*

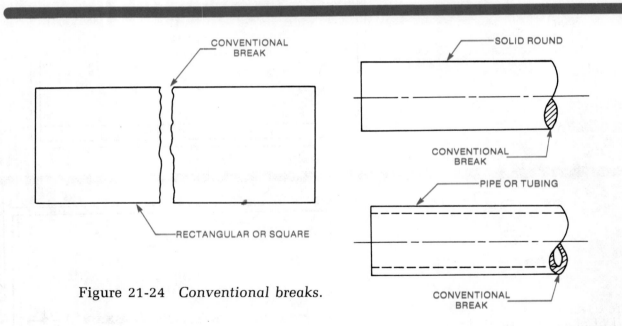

Figure 21-24   *Conventional breaks.*

conventional breaks used for solid rounds, pipe and tubing, and rectangular and square shapes. The construction steps for drawing the break for a solid round are shown in Figure 21-25. Figure 21-26 shows the construction steps for drawing the break for a pipe or tube.

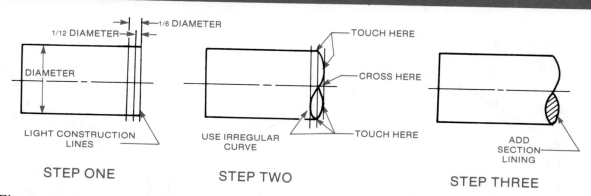

Figure 21-25 *Construction of a conventional break for solid rounds.*

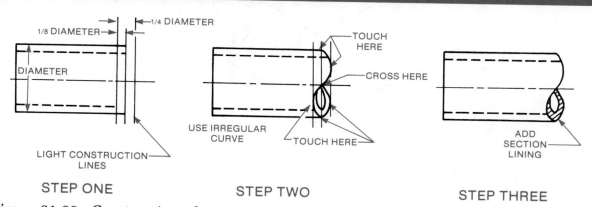

Figure 21-26 *Construction of a conventional break for pipe and tubes.*

On an A4 sheet, sketch two views of the object shown. Sketch the front view in full section. Indicate the cutting plane in the end view.

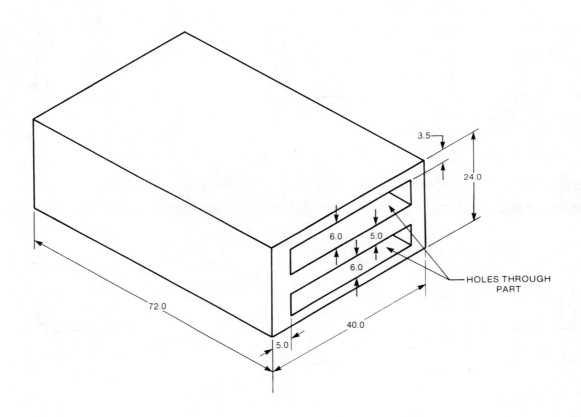

On an A4 sheet, sketch two views of the object shown. Draw the front view in half section. Indicate the cutting plane in the correct position in the side view.

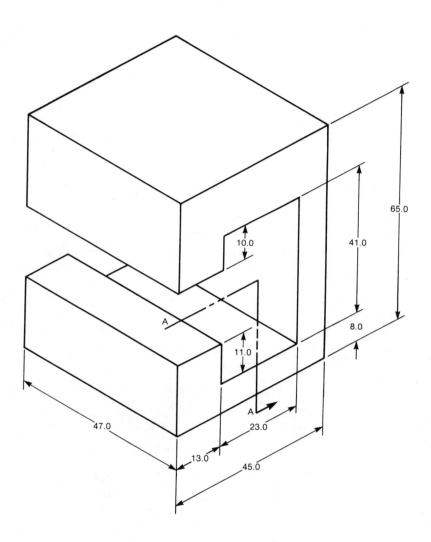

On an A4 sheet, sketch a front, top, and right side view of the object shown. In the top view, place the cutting plane for an offset section. Show the front view in section.

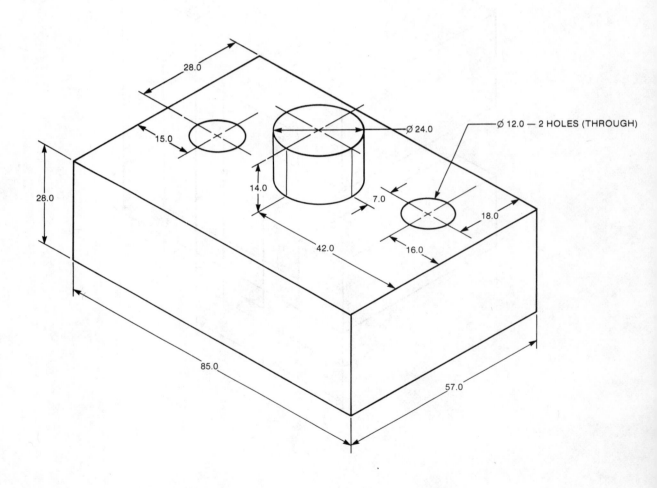

DRAWING PROBLEM 21-1 CYLINDER HOUSING

Use an A3 sheet size to prepare a multi-view drawing of the Cylinder Housing. Select a scale to suit the sheet. Draw the object with a full sectioned view. Dimension fully. Add all notes, title, and scale information.

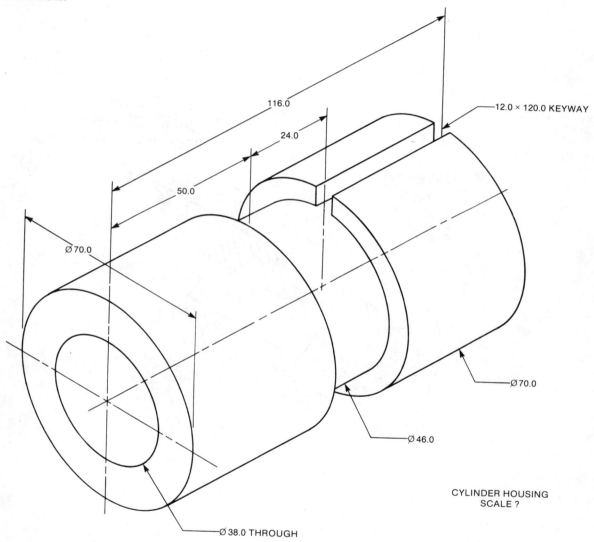

CYLINDER HOUSING
SCALE ?

On an A4 sheet prepare a multiview
drawing of the Pipe Coupling Blank.
Select a scale to suit. Draw the front view
in half section. Dimension fully. Add all
notes, title, and scale information.

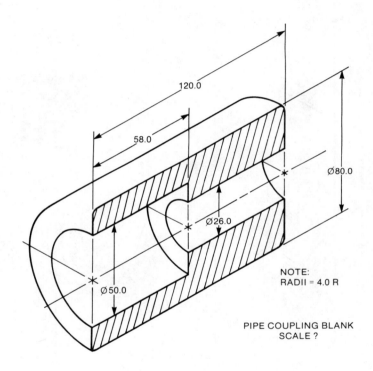

120.0

58.0

Ø80.0

Ø26.0

Ø50.0

NOTE:
RADII = 4.0 R

PIPE COUPLING BLANK
SCALE ?

DRAWING PROBLEM 21-3 CRUSHER HAMMER

Use an A2 sheet. Prepare a multiview drawing of the Crusher Hammer with a broken-out section. Select a scale to suit. Dimension fully. Add all notes, title, and scale information.

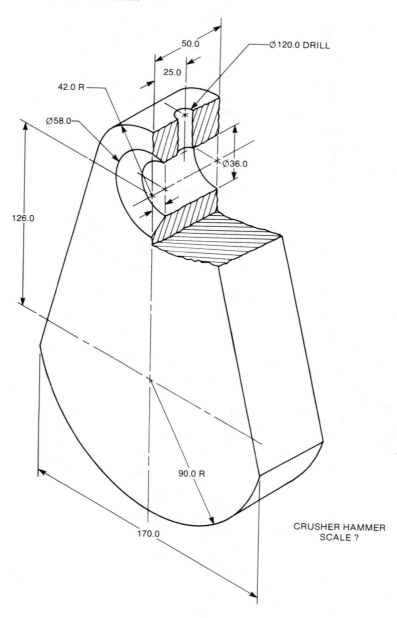

50.0

⌀120.0 DRILL

25.0

42.0 R

⌀58.0

⌀36.0

126.0

90.0 R

170.0

CRUSHER HAMMER
SCALE ?

DRAWING PROBLEM 21-4 MOTOR MOUNT

On an A3 sheet prepare a multiview drawing of the Motor Mount. Draw the object using an offset section cut at the points shown in the figure. Dimension fully. Add all notes, title, and scale information.

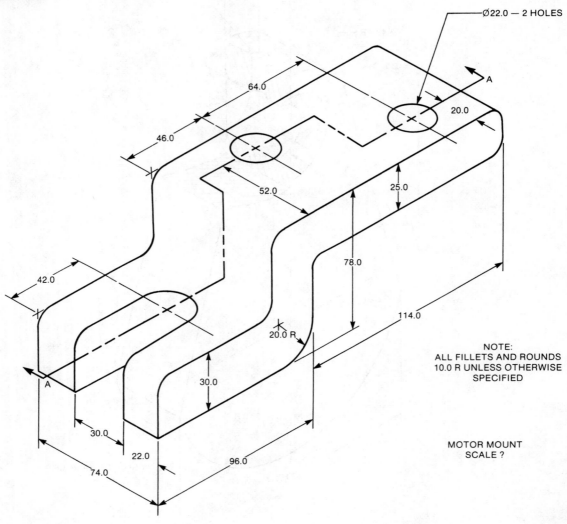

Ø22.0 — 2 HOLES

64.0

46.0

52.0

20.0

25.0

42.0

78.0

114.0

20.0 R

30.0

30.0

22.0

96.0

74.0

NOTE:
ALL FILLETS AND ROUNDS
10.0 R UNLESS OTHERWISE
SPECIFIED

MOTOR MOUNT
SCALE ?

A

DRAWING PROBLEM 21-5 OFFSET ROLLER

Use an A3 sheet. Prepare a multiview
drawing with removed sections as
indicated in the figure. Select a suitable
scale. Dimension fully. Add all notes,
title, and scale information.

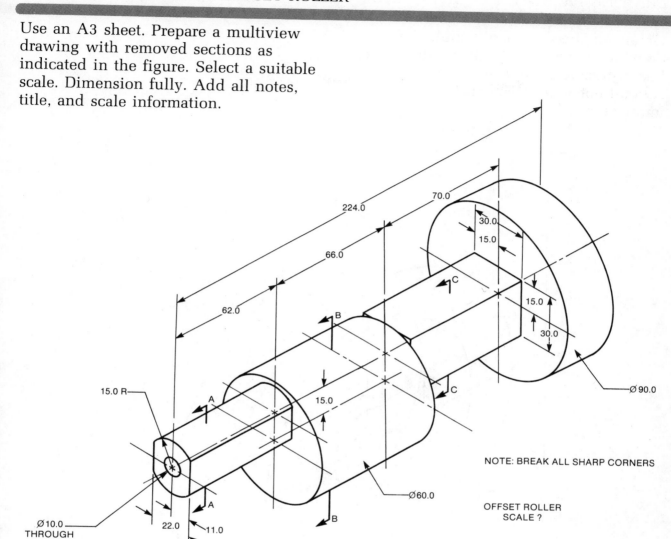

NOTE: BREAK ALL SHARP CORNERS

OFFSET ROLLER
SCALE ?

224.0

70.0

30.0

15.0

66.0

15.0

62.0

30.0

15.0 R

15.0

Ø 90.0

Ø60.0

Ø10.0
THROUGH

22.0

11.0

On an A3 sheet, prepare a multiview
drawing of the Gear Blank in full section.
Use a suitable scale. Dimension fully.
Add all notes, title, and scale
information.

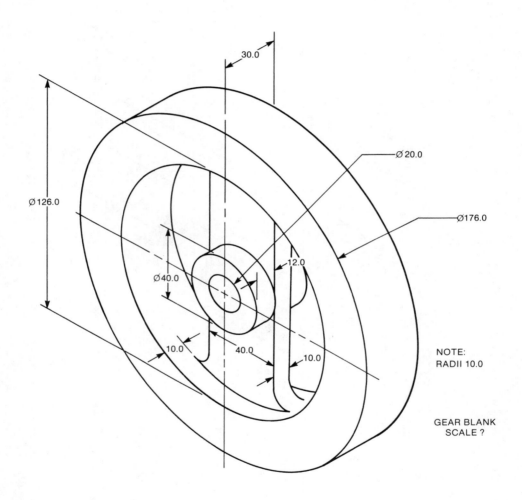

30.0

Ø 20.0

Ø126.0

Ø176.0

Ø40.0

12.0

10.0

40.0

10.0

NOTE:
RADII 10.0

GEAR BLANK
SCALE ?

DRAWING PROBLEM 21-7 PIPE NIPPLE

Use an A4 sheet. Prepare a full size, two-view drawing of the Pipe Nipple. Use a conventional break to fit the object to the sheet. Dimension fully. Add all notes, title, and scale information.

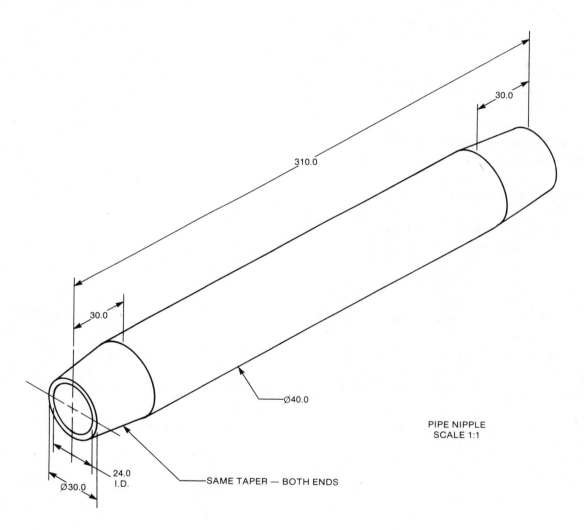

30.0

310.0

30.0

Ø40.0

PIPE NIPPLE
SCALE 1:1

24.0
I.D.

Ø30.0

SAME TAPER — BOTH ENDS

- Sectioned views show the inside features of objects.
- Sectioned views are drawn to eliminate hidden lines.
- The cutting plane indicates the point from which the section is viewed.
- Sectioned views and cutting planes are labeled with letters.
- A full section is drawn as if one-half of the object has been removed.
- Half sections are drawn as if one-fourth of the object has been removed.
- Broken-out sections are drawn as if a portion of the object has been broken away.
- Removed sections are drawn away from the view.
- The cutting plane for offset sections passes through each important feature of an object.
- Revolved sections are drawn on the view from which the section is taken.
- Section lining fills in the area "cut" in a sectioned view.
- Webs, ribs, and spokes are not section lined in sectioned views.
- Some features are rotated for sectioned views.
- Conventional breaks are used in order to fit long objects on drawing sheets.

UNIT 22

AUXILIARY VIEWS

This unit shows you how to draw objects with slanted surfaces. You will discover:

- how to draw the true form of objects with slanted surfaces,
- how to draw irregularly shaped objects clearly.

KEY WORDS

True Shape: The actual form of an object or a portion of an object (Figure 22-1).

True Size: The actual size of an object or a portion of an object (Figure 22-2).

Auxiliary View: A view on a drawing that shows the true shape and size of a surface which is not clearly shown by the regular views of a drawing.

Reference Plane: A surface used to measure auxiliary view distances.

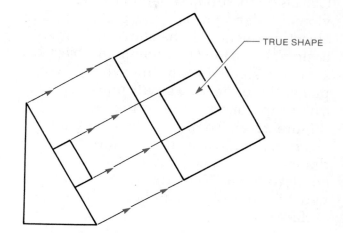

Figure 22-1   True shape.

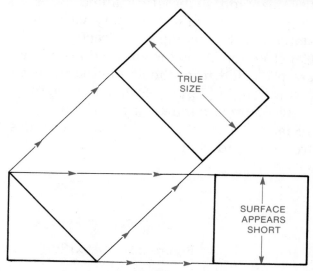

Figure 22-2   True size.

Distorted: A view of an object that shows the object not in true shape or size (Figure 22-3).

OBJECTS WITH SLANTED SURFACES

Objects with surfaces parallel to the viewing plane present few problems. The object can generally be shown in three basic views. However, when an object has a surface slanted away from the viewing plane, the basic views will not show the *true shape* or *true size* of the object (Figure 22-4). To aid in describing an object not shown in its true form, a drafter uses a special view called an *auxiliary view*. The auxiliary view adds a new viewing plane parallel to the slanted surface.

THE AUXILIARY VIEW

An auxiliary view is drawn to show the true size and shape of a slanted surface (Figure 22-5). On an auxiliary view the slanted surface is viewed directly. Construction lines for an auxiliary view are projected from the slanted surface on a right angle (90 degrees). The imaginary view plane is parallel to the slanted surface. An auxiliary view, therefore, has true size and shape.

Figure 22-5 *Auxiliary view.*

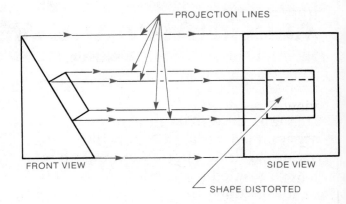

Figure 22-3 *A distorted shape.*

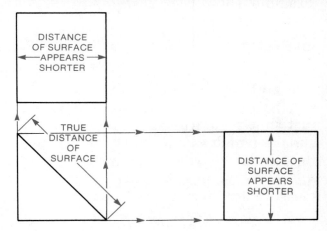

Figure 22-4
Some views may not show true shape or true size.

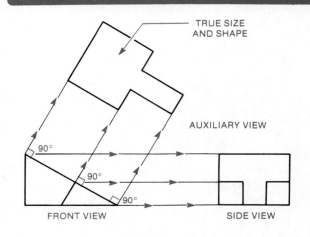

REFERENCE PLANE

A reference plane refers to a side of the surface that gives a key measurement to the auxiliary view. The reference plane is used to establish a side of the auxiliary view (Figure 22-6).

TYPES OF AUXILIARY VIEWS

There are three types of auxiliary views: length auxiliary, height auxiliary, and width auxiliary. Each auxiliary view takes its name from the major dimension shown in the view. Figure 22-7, for example, shows a typical width auxiliary view.

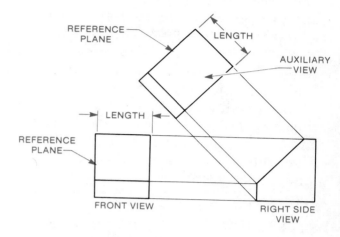

Figure 22-6 *Width auxiliary view.*

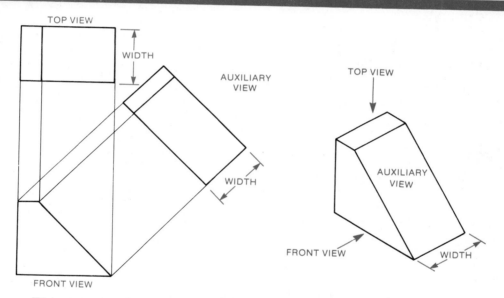

Figure 22-7 *Drawing the projection lines and reference plane.*

The Length Auxiliary View. The length auxiliary view is projected at 90 degree angles from the right side view (Figure 22-8). The length auxiliary view is completed by transferring the length from the front view (Figure 22-6).

The length auxiliary view is drawn in this way. For projection, place a T-square and triangle as shown in Figure 22-9. Then project the distances from the side view at 90 degrees to the slanted surface.

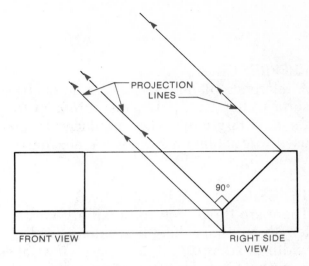

Figure 22-8   *Projecting a length auxiliary view.*

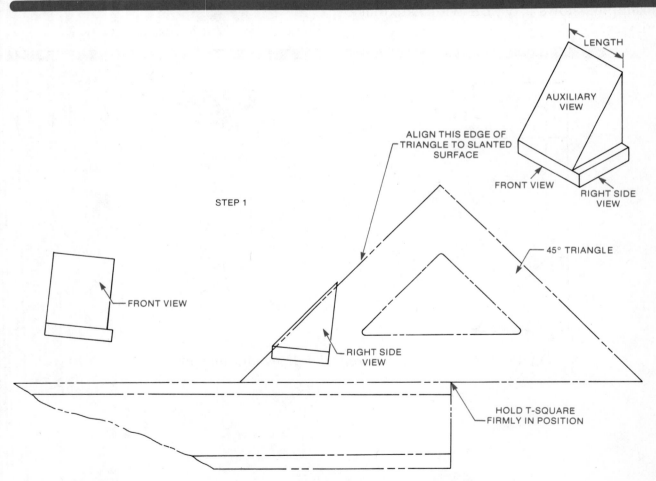

Figure 22-9   *The length transferred.*

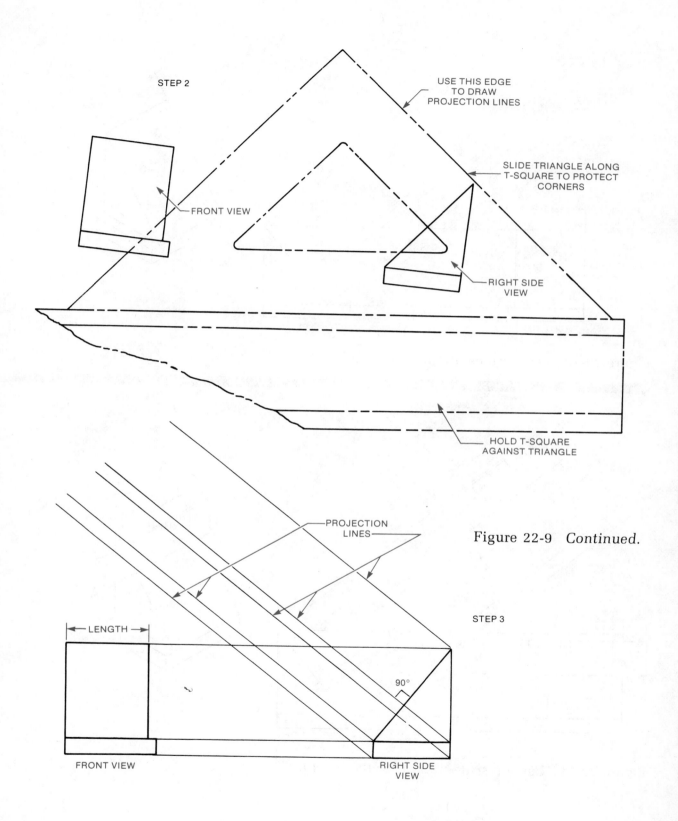

STEP 2

USE THIS EDGE
TO DRAW
PROJECTION LINES

FRONT VIEW

SLIDE TRIANGLE ALONG
T-SQUARE TO PROTECT
CORNERS

RIGHT SIDE
VIEW

HOLD T-SQUARE
AGAINST TRIANGLE

PROJECTION
LINES

Figure 22-9 Continued.

STEP 3

LENGTH

90°

FRONT VIEW

RIGHT SIDE
VIEW

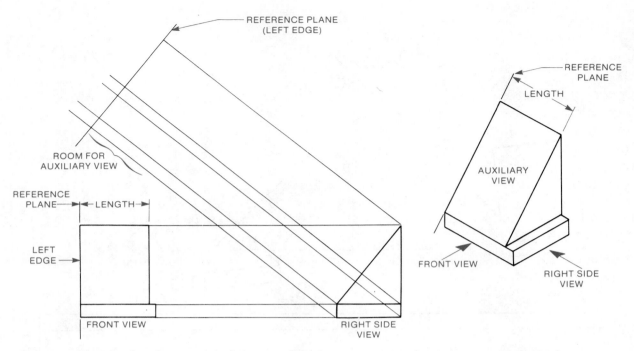

Figure 22-10   *Left edge as a reference plane.*

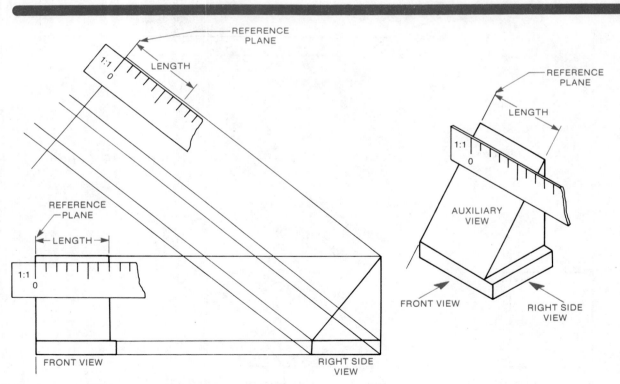

Figure 22-11   *Using a scale to transfer the length.*

Draw a reference plane parallel to the slanted surface (Figure 22-10). Place the reference plane so that there will be enough room for the entire auxiliary view. The reference plane will represent the object's left edge as seen in the front view.

Next, transfer the length distances from the front view. These distances can be transferred with a scale (Figure 22-11) or dividers (Figure 22-12).

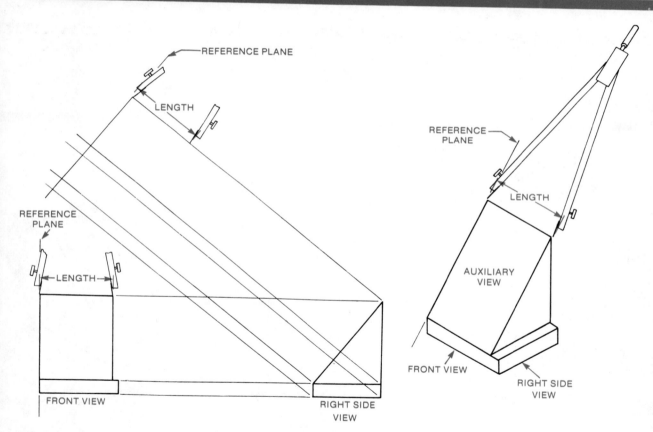

Figure 22-12   *Using a divider to transfer the length.*

To avoid confusion, number all the corners of the object. This will allow you to identify (with numbers) the corners in the auxiliary view (Figure 22-13).

To complete the view, close off each edge (Figure 22-14).

The Height Auxiliary View. The height auxiliary view is drawn by projecting distances from the top view and transferring distances from the side view (Figure 22-15). As with the length auxiliary view, a reference plane should be used.

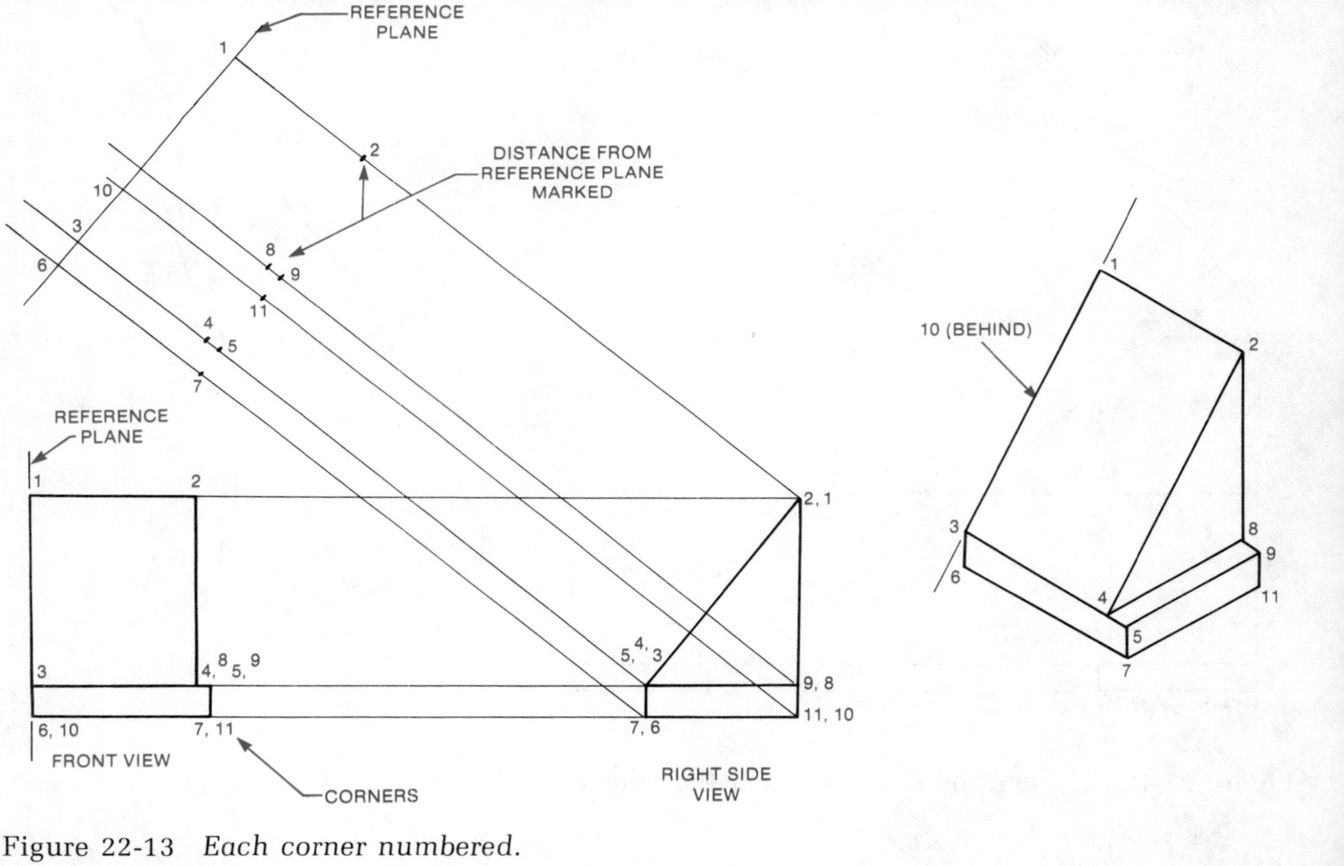

Figure 22-13 *Each corner numbered.*

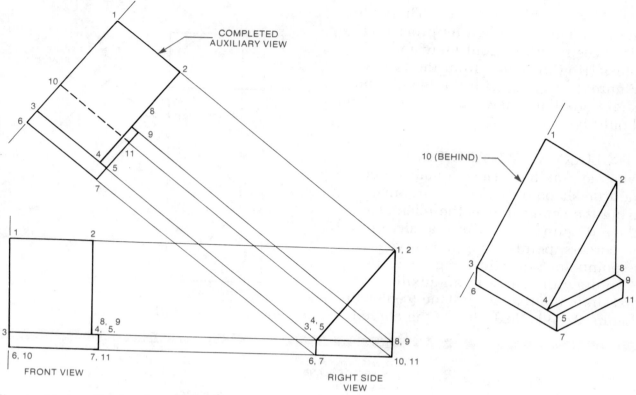

Figure 22-14   *Each edge closed off.*

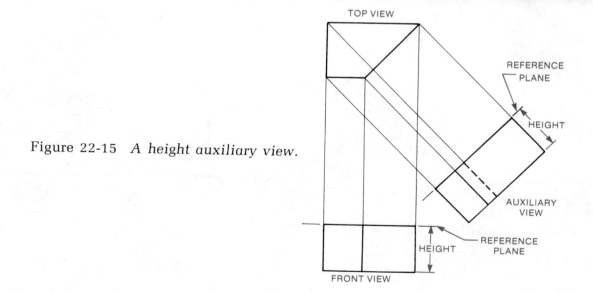

Figure 22-15   *A height auxiliary view.*

The Width Auxiliary View. The width auxiliary view is drawn by projecting distances from the front view and transferring distances from the top view (Figure 22-16). As with the length and height auxiliary views, a reference plane should be used.

AUXILIARY VIEWS AND CURVES

When an auxiliary view contains a curve, the true shape of the curve is drawn using the center line as the reference plane (Figure 22-17). Since a curve has no corners, points along the curve serve as points for projection (Figure 22-18).

To draw the curve in an auxiliary view, first draw the reference plane (center line) (Figure 22-19). Next, divide the curved

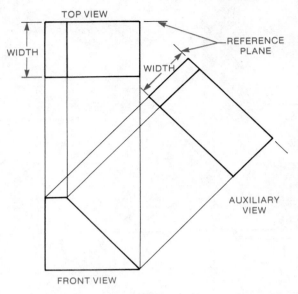

Figure 22-16   *A width auxiliary view.*

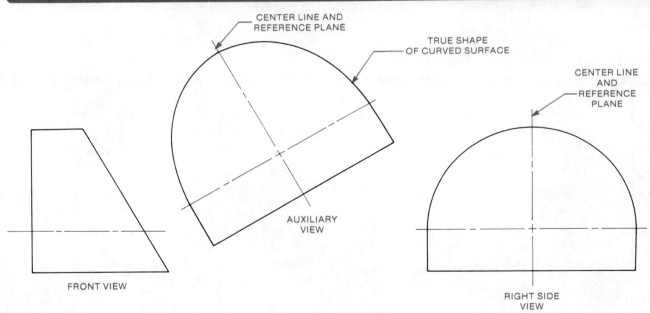

Figure 22-17   *An auxiliary view of a curved surface.*

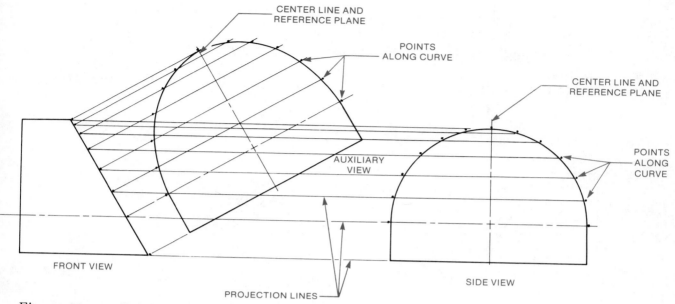

Figure 22-18 Points along the curve are used to project distances.

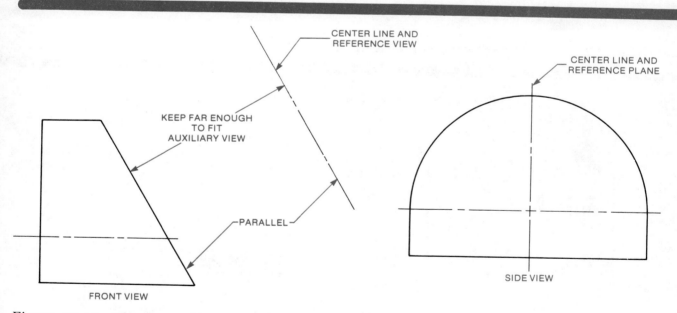

Figure 22-19 The center line is used as the reference plane.

surface into 15 degree angles to locate points along the curve view (Figure 22-20). It is helpful to lightly number the points (Figure 22-21).

Next, project the points to the slanted surface and then to the auxiliary view (Figure 22-22). Then transfer the distances from the center line (reference plane) to the auxiliary view (Figure 22-23). To connect the points in the auxil-

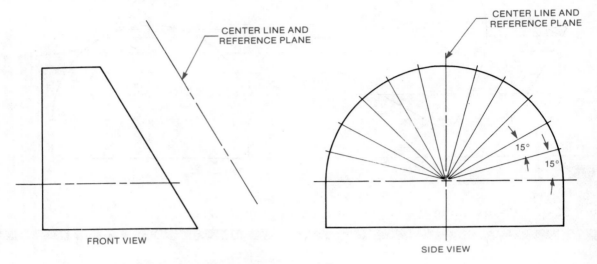

Figure 22-20 *The curved surface is divided.*

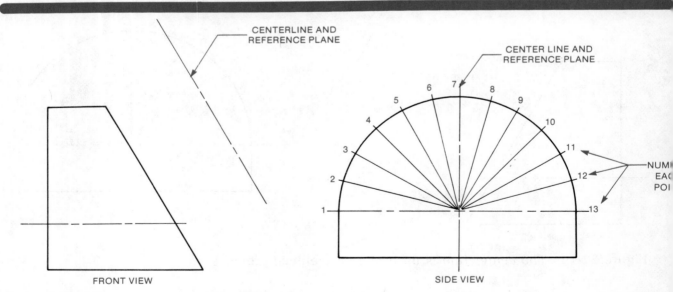

Figure 22-21 *Points are numbered.*

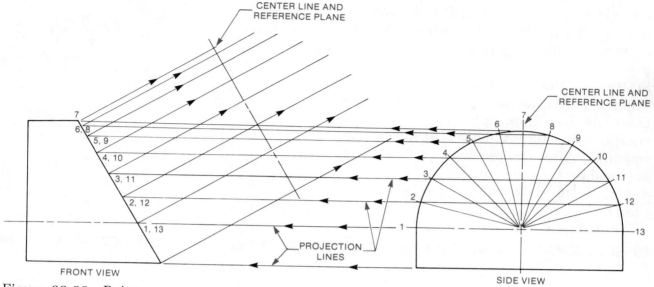

Figure 22-22 Points are projected.

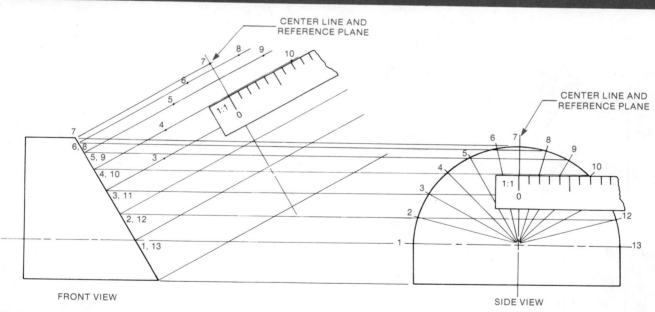

Figure 22-23 Distances transferred from the reference plane.

iary view, draw a light freehand curve (Figure 22-24). Use an irregular curve to darken the line (Figure 22-25).

SIMPLIFIED (PARTIAL) VIEWS
When a partial auxiliary view is used, the complete front, top, right side, or complete auxiliary view might not be required. When one or more of these views shows a portion of the object *distorted*, the distorted portion is not drawn (Figure 22-26). The view is closed off by using an irregular break line.

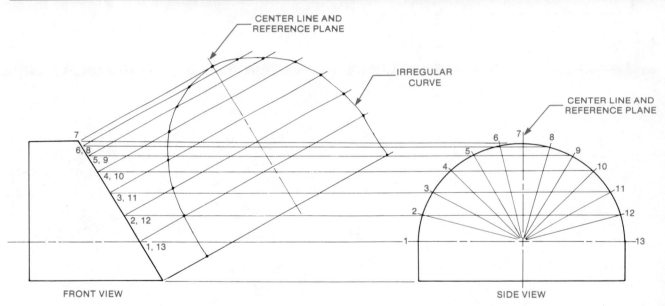

Figure 22-24 *Points are first connected by a freehand curve.*

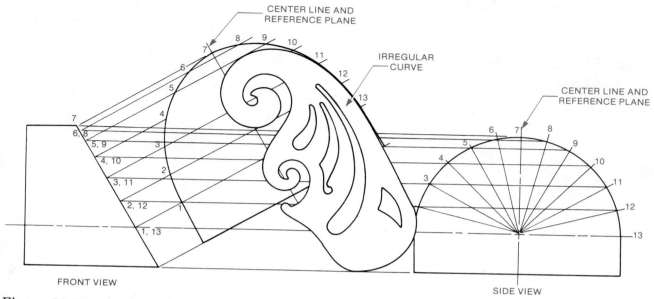

Figure 22-25   An irregular curve is used to darken the curve.

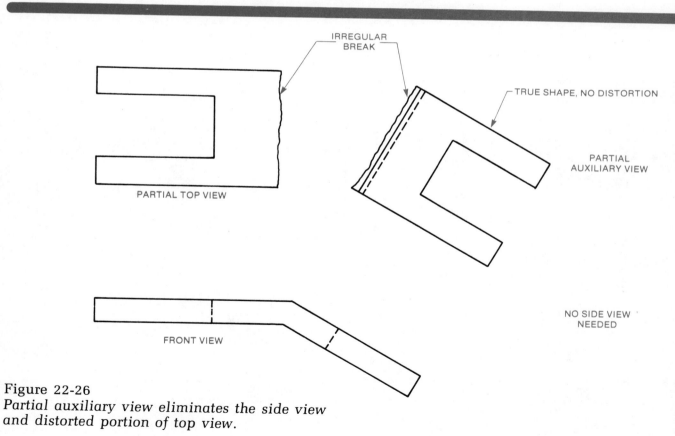

Figure 22-26
Partial auxiliary view eliminates the side view
and distorted portion of top view.

On an A4 sheet, sketch or draw the front, top, and right side views as shown. Next, make a length auxiliary view by projecting the distances from the right side view.

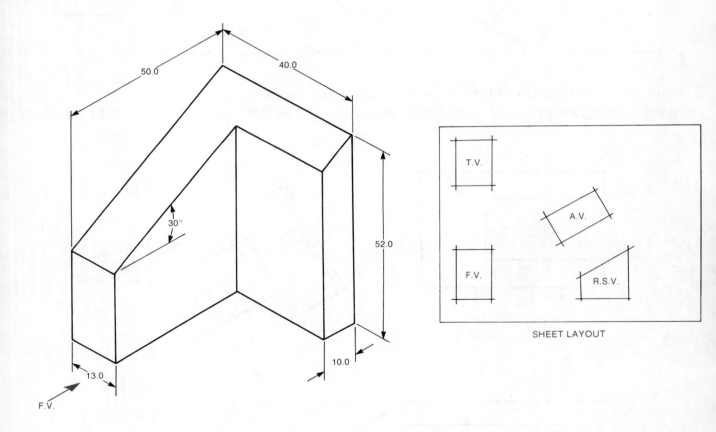

SHEET LAYOUT

On an A4 sheet, sketch or draw the front, top, and right side views as shown. Next, make a height auxiliary view by projecting the distances from the top view.

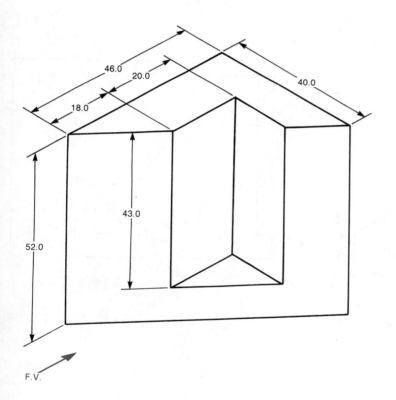

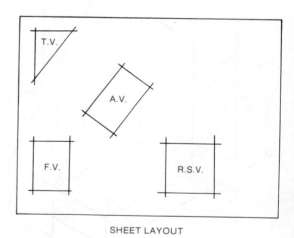

SHEET LAYOUT

On an A4 sheet, sketch or draw the front,
top, and right side views as shown. Next,
make a width auxiliary view by projecting the distances from the front view.

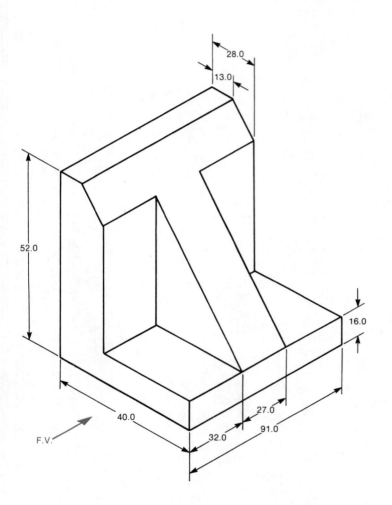

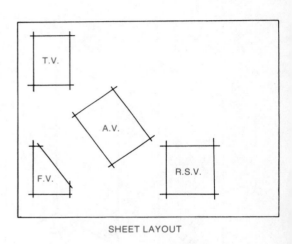

SHEET LAYOUT

Use an A3 sheet size. Prepare a multi-view drawing of the Mounting Bracket. Include partial views and auxiliary views. Select a scale to suit. Dimension fully. Add all notes, title, and scale information.

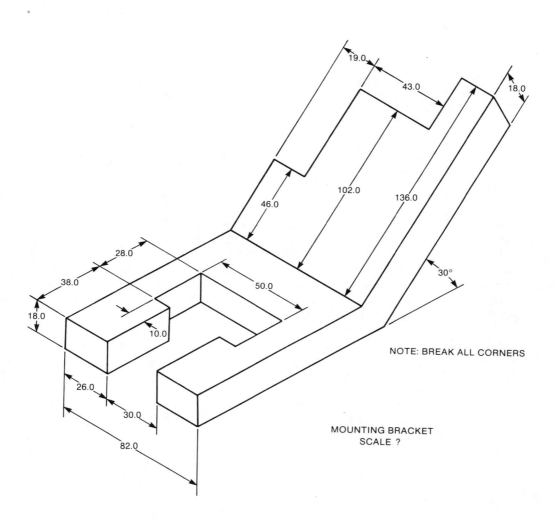

NOTE: BREAK ALL CORNERS

MOUNTING BRACKET
SCALE ?

Use an A3 sheet. Prepare a multiview
drawing with auxiliary view of the
Positioning Block. Select a suitable scale.
Dimension fully. Add all notes, title, and
scale information.

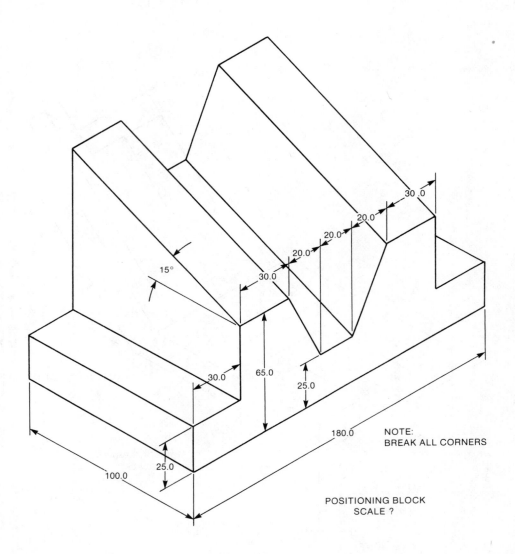

15°

30.0

30.0

20.0

20.0

20.0

30.0

65.0

25.0

180.0

NOTE:
BREAK ALL CORNERS

100.0

25.0

POSITIONING BLOCK
SCALE ?

On an A3 sheet prepare a multiview
drawing with a partial auxiliary view of
the Cover Strap. Draw to a suitable scale.
Dimension fully. Add all notes, title, and
scale information.

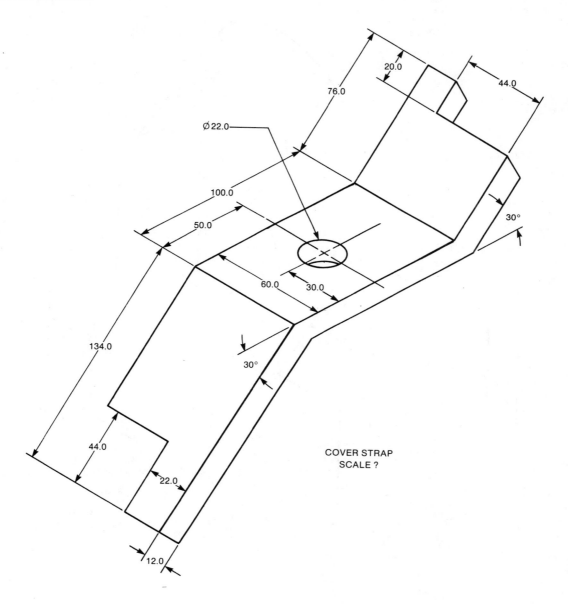

COVER STRAP
SCALE ?

Using an A4 sheet, prepare a fully
dimensioned multiview drawing of the
Pipe Support. Select the best views. Add
all notes, title, and scale information.

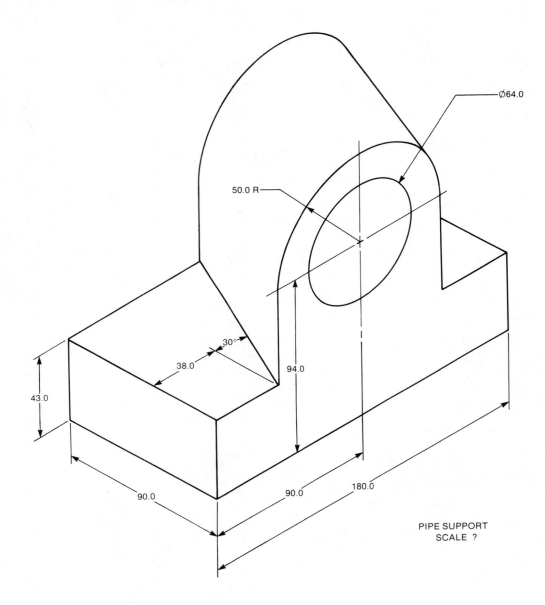

Ø64.0

50.0 R

30°

38.0

94.0

43.0

90.0

90.0

180.0

PIPE SUPPORT
SCALE ?

On an A3 sheet draw a full scale (1:1)
multiview drawing of the Fuel Line Tube.
Dimension fully. Add the title, scale, and
all notes.

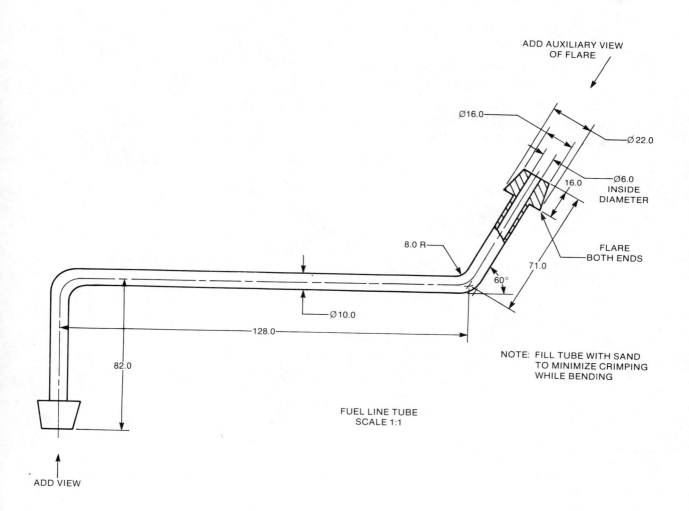

ADD AUXILIARY VIEW
OF FLARE

Ø16.0

Ø 22.0

Ø6.0
INSIDE
DIAMETER

16.0

FLARE
BOTH ENDS

8.0 R

71.0

60°

Ø10.0

128.0

82.0

NOTE: FILL TUBE WITH SAND
TO MINIMIZE CRIMPING
WHILE BENDING

FUEL LINE TUBE
SCALE 1:1

ADD VIEW

- Auxiliary views are drawn to show the true shape and size that cannot be seen in the basic views.
- An auxiliary view is drawn looking directly at the slanted surface.
- The three types of auxiliary views are: length, width, and height auxiliary views.
- A reference plane is used to transfer distances for auxiliary views.
- Curves in auxiliary views are drawn by projecting points located on the curve and then connecting them.
- Simplified (partial) views are used instead of drawing distorted portions of views.

DRAFTING MATH

This unit shows you the basic math needed in the drafting room. You will:

- review the basic math skills that will be needed for drafting,
- review how to add, subtract, multiply, and divide decimals correctly.

KEY WORDS

Decimal: A portion of a whole number (Figure 1).

Decimal Point: A dot or period that is put to the left of a decimal (Figure 1).

Digit: Any number 0 through 9.

MATH AND DRAWING

To draw accurately, you must measure accurately. When you measure, it often becomes necessary to add, subtract, multiply, and divide to determine the distances to be measured. Your ability to perform math problems has a direct bearing on the accuracy of your drawings. This unit will serve as a review so you can sharpen your math skills.

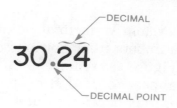

Figure 1   *A decimal and a decimal point.*

DECIMAL NUMBERS AND DRAFTING

Since the metric system is used for measuring in the drafting room and the metric system is a *decimal* based system, the drafter must be skilled in the use of decimals.

In the decimal system, a *decimal point* separates whole numbers from portions of whole numbers (Figure 2). Whole numbers always appear to the left of the decimal point. Portions of whole numbers always appear to the right of the decimal point.

Whole numbers in decimal form are no different than whole numbers that appear without decimal points. For example:

$$55.0 = 55$$
$$107.00 = 107$$

Numbers that appear in the first place to the right of the decimal point are actually tenths of the whole number *one*: for example, .5 is five tenths (5/10).

Numbers that appear in the second place to the right of the decimal point are hundredths of the whole number *one*: for example, .05 is five hundredths (5/100).

Numbers that appear in the third place to the right of the decimal point are thousandths of the whole number *one*: for example, .005 is five thousandths (5/1000).

ADDING DECIMALS

When adding decimals, it is important that you keep the decimal points directly below decimal points and *digits* aligned with the digits above and below (Figure 3). Each column is added beginning at the extreme right-hand column.

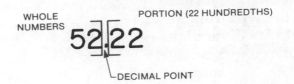

Figure 2
The decimal point separates the whole number from the part of a whole.

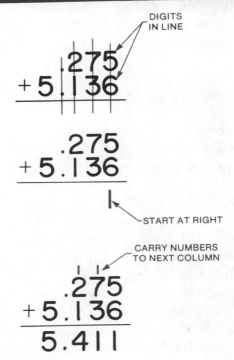

Figure 3
Top, *lining up digits for addition;* center, *begin adding from the right;* bottom, *carrying the numbers to the next column.*

Add the following groups of decimals. Be sure to keep the digits in line and the decimal points below each other. Write problems and answers on a separate sheet of paper.

1) 1.125 + 3.462

2) 1.062 + 5.7 + 8.469

3) 19.25 + .006 + .72 + .005

4) 207.5 + 1900.0 + .007

5) 3.2 . + 11.005 + .06) + .094 + 8.2611

6) 4 and 32 hundredths plus 1 and 2 tenths

7) One hundred six and four tenths plus thirty-two hundredths.

SUBTRACTING DECIMALS

As with adding, it is important to keep the decimal points and digits directly below each other when subtracting (Figure 4). Begin with the right-hand column and work toward the left. It may become necessary to "borrow" from a digit to the left of the column you are subtracting. When borrowing, be sure to cancel one from the column on the left and add ten to the column on the right.

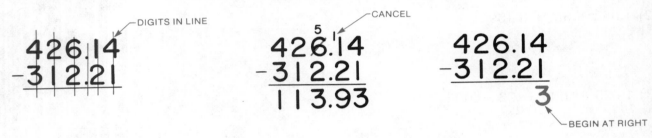

Figure 4
Line up digits for subtraction. Begin by subtracting from the right. Borrow and cancel.

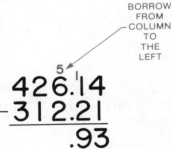

Subtract the following groups of decimals.

1) .05 − .03

2) .065 − .057

3) 7.162 − 1.032

4) 141.898 − 120.007

5) 62.6591 − 61.7284

6) Forty-nine and three tenths minus sixteen and fifty-four hundreths

7) Seventy-eight hundreths minus six tenths

MULTIPLYING DECIMALS

Figure 5, *left*, shows whole number multiplication. When you multiply decimals, count the numbers to the right of the decimal point in both numbers. In Figure 5, *right*, for example, there are two places. After multiplying the numbers count the same number of total places starting from the right (in this case, two). Place the decimal point.

Since both numbers multiplied in Figure 6 have a total of five places, the decimal point in the answer is placed five places to the left of the last digit.

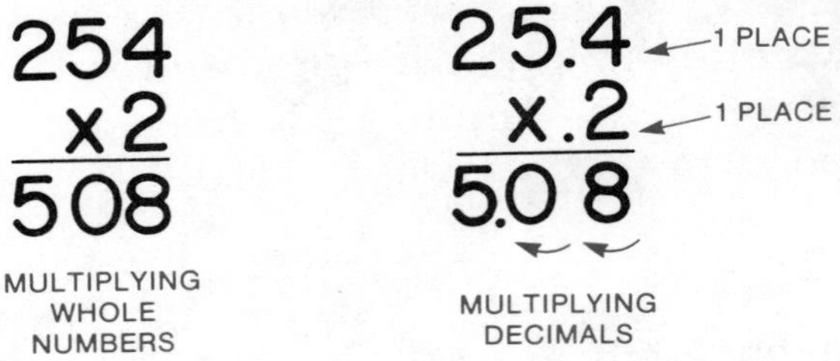

$$\begin{array}{r} 254 \\ \times 2 \\ \hline 508 \end{array}$$

MULTIPLYING
WHOLE
NUMBERS

$$\begin{array}{r} 25.4 \quad \leftarrow 1\ \text{PLACE} \\ \times .2 \quad \leftarrow 1\ \text{PLACE} \\ \hline 5.08 \end{array}$$

MULTIPLYING
DECIMALS

Figure 5
Multiplying whole numbers and decimals.

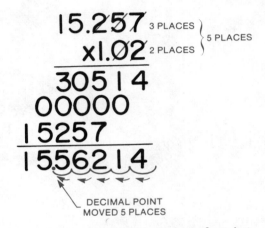

15.257 3 PLACES ⎫
x1.02   2 PLACES ⎬ 5 PLACES
                 ⎭

30514
00000
15257
15562 14

DECIMAL POINT
MOVED 5 PLACES

Figure 6   *Placing the decimal point.*

Multiply the following groups of
decimals. Be sure to place the decimal
point in your answer correctly.

1) .25 × .17

2) .350 × .658

3) 1.437 × 5.625

4) 18.05 × .006

5) 191.0 × .566

6) Seventy-one thousandths times five hundreths

7) Thirteen and sixty-four hundreths times five tenths

DIVIDING DECIMALS

Dividing decimals is similar to dividing whole numbers. The difference is the placement of the decimal point in the answer. You must first count the number of digits to the right of the decimal point in the number by which you are dividing (divisor) (Figure 7). Move the decimal that same number of places to the right in the number you will be dividing into (dividend). Divide as you would divide whole numbers.

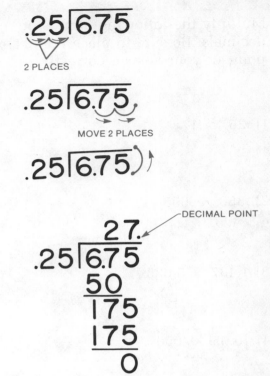

Figure 7
Count the number of digits in the divisor. Move the decimal point in the dividend. Place the decimal point in the quotient.

Divide the following groups of decimals.
Take care in first locating the decimal
point before dividing.

1) $.25 \overline{)3.75}$

2) $.002 \overline{)9.328}$

3) $6.32 \div .015$

4) $18.731 \div 4.3$

5) $1.726 \div 1.32$

6) Thirty and ninety-one hundreths divided by eleven hundreths

7) Four tenths divided by eight hundreths

- The decimal point is to the left of the decimal digits.
- Whole numbers are always shown to the left of the decimal point.
- Portions of whole numbers are to the right of the decimal point.
- The decimal places from the decimal point to the right are called tenths, hundredths, thousandths, and ten thousandths.
- It is important to locate the decimal point correctly when adding, subtracting, multiplying, and dividing decimals.

COMPETENCY TESTS

UNIT 1 COMPETENCY TEST

Write answers on a separate sheet of paper.

1) Why were some of the first methods of measurements inaccurate?

2) How did the metric system improve accuracy?

3) What does S.I. stand for?

4) What unit is used for length measurement on mechanical drawings?

5) How is the millimetre derived from the metre?

UNIT 2 COMPETENCY TEST

Write answers on a separate sheet of paper.

1) How do prehistoric drawings compare with the drawings industry uses today?

2) Why is drafting called a "Graphic Language"?

3) Why was drafting important during the Industrial Revolution?

4) List some complicated items in which drafting played an important role.

5) What is the difference between a chief drafter's job and a group leader's job?

6) What does a designer do?

7) What two people does a junior drafter help?

8) Why does a checker need to have much drafting knowledge?

9) What does a reproduction technician do?

UNIT 3 COMPETENCY TEST

Write answers on a separate sheet of paper.

1) What determines the type of drafting board you buy?

2) The length of the T-square should suit the _____.

3) What two triangles are needed?

4) How are pencils graded?

5) What basic equipment does a drafting machine replace?

6) Why should wooden items never be cleaned with water?

7) How should a sandpaper pad be stored?

UNIT 4 COMPETENCY TEST

Write answers on a separate sheet of paper.

1) What levers are released on a drafting machine before you begin drawing?

2) How are scales removed from a drafting machine?

3) The protractor head stops at every _____ degree mark.

4) How is an angle of seven degrees held on a drafting machine?

5) How should a drafting machine be cleaned?

6-10) Identify the numbered parts of the drafting machine at the right.

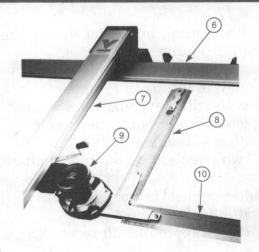

UNIT 5 COMPETENCY TEST

Write answers on a separate sheet of paper.
1) Which pencil lead is harder, 2H or H?
2) What four grades of pencils should be used on vellum and cross section paper?
3) Which pencil should be used for thick dark lines? For thin dark lines?
4) Which pencil should be used for lettering?
5) Why should pencil pointing be done away from your work?
6) Pencils are pointed using a _____ or a _____.

UNIT 6 COMPETENCY TEST

Write answers on a separate sheet of paper.
1) What pencil grades are best for sketching?
2) How can sketching help plan a drawing?
3) Name the two types of oblique sketches.
4) Why is one oblique sketching type used more than the other?
5) How does isometric sketching differ from oblique sketching?
6-7) Identify these two forms of sketching.

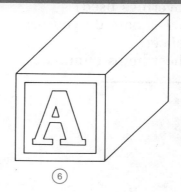

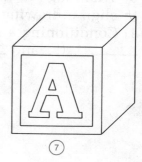

UNIT 7 COMPETENCY TEST

Write answers on a separate sheet of paper.
1) What are the main features of vellum?
2) Why is film better than vellum?
3) Cross section paper is used for _____ and _____.
4) What is the smallest metric sheet size? the largest?
5) How does an A4 sheet compare to an A1 sheet?
6) If an A3 sheet was cut in half, what size sheets would you have?
7) List the items found in a title block.
8-11) Identify the sheet sizes shown at the right.

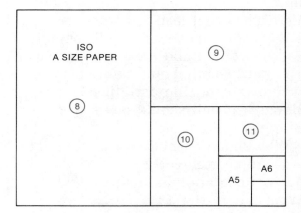

UNIT 8 COMPETENCY TEST

Write answers on a separate sheet of paper.
1) Name the three types of board coverings.
2) Why are board coverings used?
3) What edge of the T-square do you use to align a drawing sheet?
4) Conditioning a sheet helps eliminate _____ and _____.

290   DRAFTING: METRIC

Write answers on a separate sheet of paper.

1) *Vertical* means
  A. Sideways
  B. Up and down
  C. Parallel
  D. Slanted

2) *Horizontal* means
  A. Sideways
  B. Up and down
  C. Parallel
  D. Slanted

3) What preparation is required before starting a drawing?

4) On which edge of your drawing board should the T-square head be? Why?

5) Which edge of the T-square is used for drawing lines?

6) In which two directions should you tilt your pencil?

7) Explain why a pencil should be rotated.

8) How would you draw a vertical line with pencil, T-square, and triangle?

9) How should you move a triangle? Why?

10) How often should your pencil be pointed?

11-12) Identify the lines at the right.

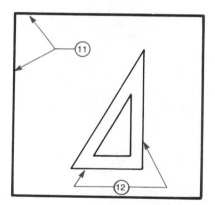

Write answers on a separate sheet of paper.

1) Most inclined lines in mechanical drawing are at _____ degree increments.

2) It is possible to draw 45 degree lines inclined in _____ directions.

3) How many degrees does angle A have?

4) How many degrees does angle B have?

5) How many degrees does angle C have?

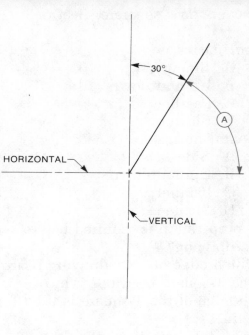

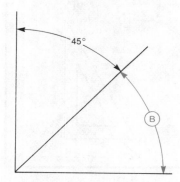

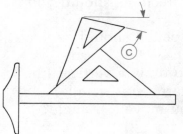

UNIT 11 COMPETENCY TEST

Write answers on a separate sheet of paper.

1) Why is the scale marked 1:1 called a full scale?

2) All measurements made with the full scale should be started at which mark?

3) Which markings on the full scale are longer than the single millimetre marks? Why?

4) Where are half millimetre distances found on the full scale?

5-10) Indicate the distances from zero marked on the scale shown here.

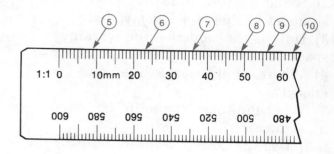

UNIT 12 COMPETENCY TEST

Write answers on a separate sheet of paper.

1) What style of lettering is used on mechanical drawings?

2) Which letters are four units wide?

3) Which letters are wider than four units?

4) How wide should the spacing between words be?

5) What instruments are used to make guidelines?

Write answers on a separate sheet of paper.

1) Objects ——————— thick or less are usually drawn with the thickness omitted.

2) Name the four lines included in the alphabet of lines for flat layouts.

3) How is the center of the working space found?

4) When should excess line work be erased?

5) What information should be lettered on a drawing?

6-8) Use the correct term to describe the distances shown at the right.

9-11) Identify the lines at the right.

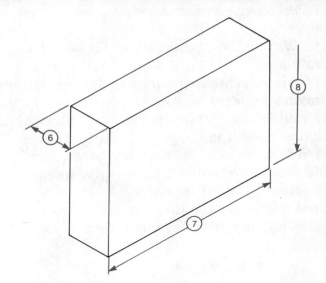

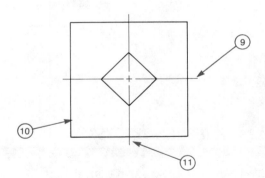

UNIT 14 COMPETENCY TEST

Write answers on a separate sheet of paper.

1) What is the radius of a 14 mm circle?

2) What is the diameter of a circle with a 15 mm radius?

3) Why should a compass be rotated in one direction only?

4) Describe a fillet.

5) How does a round differ from a fillet?

6) What is a tangency point?

7-8) Identify the arcs shown here.

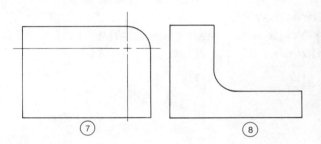

Write answers on a separate sheet of paper.

1) What two methods can be used to draw a square?

2) How can an equilateral triangle be drawn?

3) Which triangle is needed to draw a hexagon?

4) A hexagon can be drawn if two distances are known. Which distances?

5) When drawing an octagon you will need which triangle?

6-10) Identify the shapes shown below.

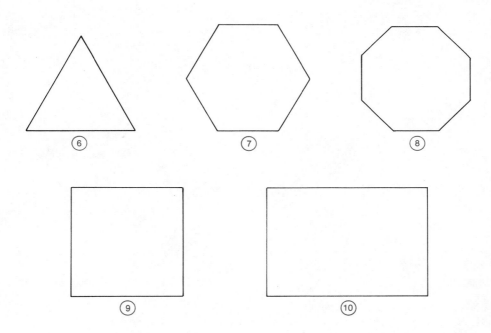

Write answers on a separate sheet of
paper.
1) Describe the distance measured when
measuring an angle.
2) What is the point where two sides of
an angle meet?
3) What must be aligned when using the
protractor to measure angles?
4) Describe an acute angle.
5) Describe a reflex angle.
6-10) Identify the angles shown here.

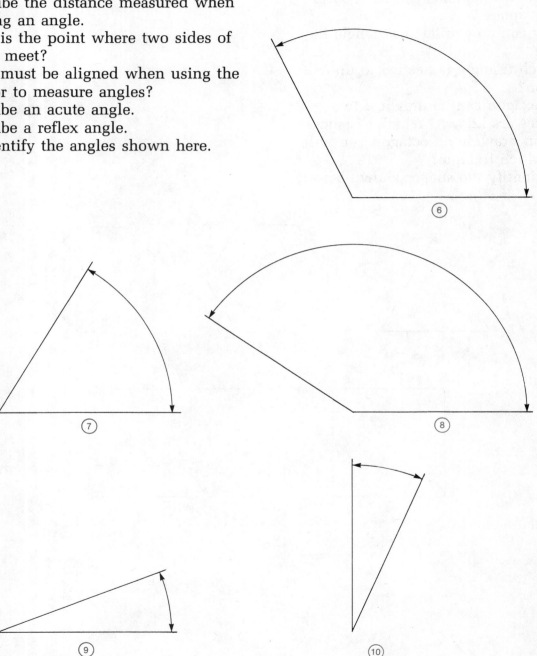

UNIT 17 COMPETENCY TEST

Write answers on a separate sheet of paper.

1) What three views are the most common?

2) What are the alternate views?

3) What is a miter line?

4) What dimensions (length, width, height) do the top and front views have in common?

5) What two dimensions do the front and side views have in common?

6) Where in the drawing is the front view located when the top and right side views are also used?

7-11) Study the top figure. Indicate on which view the numbered surfaces would appear (front, top, right side).

12-15) Identify the lines in the bottom figure shown here.

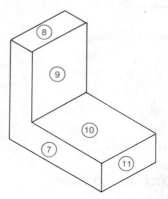

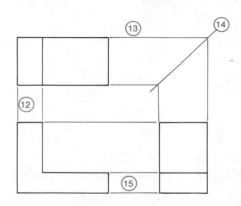

UNIT 18 COMPETENCY TEST

Write answers on a separate sheet of paper.

1) When should hidden lines be used?

2) How long should the dashes of a hidden line be?

3) When a hidden line and object line are to occupy the same space, which line should be omitted?

4) Sketch a hidden line crossing an object line.

5) Sketch two hidden lines meeting.

UNIT 19 COMPETENCY TEST

Write answers on a separate sheet of
paper.
1) Name three types of lines used for
dimensioning.
2) Why should dimensions be placed
away from a view when possible?
3) What size should the numbers and
letters for dimensions and notes be?
4) How far from the object should the
first dimension line be?
5) What is the symbol for diameter? for
radius?
6) What are some things to consider
before placing dimensions?
7-10) Identify the numbered items at the
right.

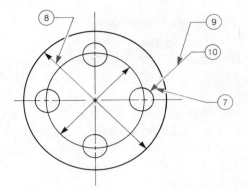

UNIT 20 COMPETENCY TEST

Write answers on a separate sheet of
paper.
1) When should a reduction proportion
scale be used?
2) When is an enlargement proportion
scale used?
3) What are the most common reduction
proportion scales?
4) When a 1:10 proportion scale is used,
how many millimetres are drawn for
every 10 millimetres of the actual object?
5) How many millimetres does each mark
represent when the full scale is used for
a 2:1 scale drawing?
6-10) Identify the distances from zero
shown here.

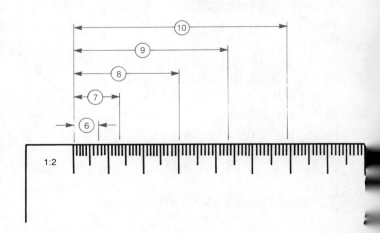

298   DRAFTING: METRIC

Write answers on a separate sheet of paper.
1) What are some reasons for using sectioned views?
2) List the types of sectioned views.
3) Which line is used to indicate the point where a section is taken?
4) What determines the spacing of section lining?
5) Which features are drawn without section lining in a sectioned view?
6) Why are some features revolved for sectioned views?
7) Identify the three types of sections shown here.

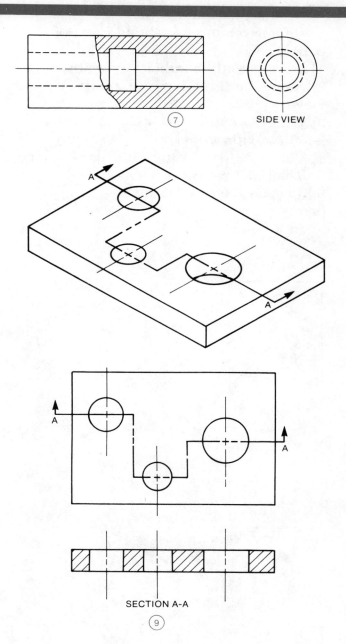

SIDE VIEW

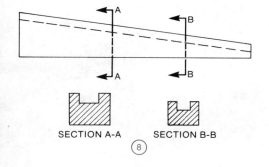

SECTION A-A   SECTION B-B

⑧

SECTION A-A

⑨

Write answers on a separate sheet of paper.

1) When should auxiliary views by used?

2) Name the three types of auxiliary views.

3) What two views are used to draw a length auxiliary view?

4) Why would a partial view be drawn?

5) What is a reference plane?

6-8) Name the auxiliary views shown here.

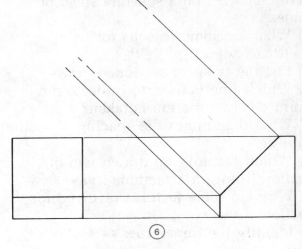

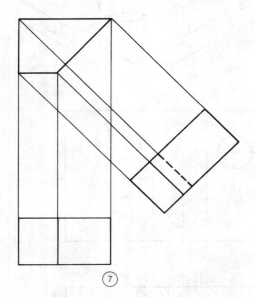

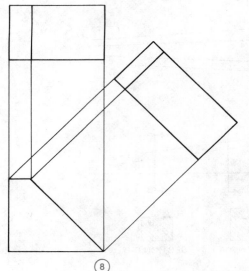

Write answers on separate sheet of paper.

1) Numbers to the left of the decimal point are _____ numbers.

2) Numbers to the right of the decimal point are _____ of whole numbers.

3) Write "three hundred twenty-seven thousandths" in decimal form.

4) Add the following decimals: $.75 + 1.06 + 2.032$.

5) Subtract 3.349 from 7.285.

6) Multiply $.037 \times 5.22$.

7) Divide 807.66 by 47.3.

INDEX